THE PROFESSIONAL ENGINEERING CAREER DEVELOPMENT SERIES

ENGINEERING PROFESSION ADVISORY GROUP

Dr. Joseph M. Biedenbach, Director, Engineering Education Programs, RCA

Dean R. L. Bisplinghoff, School of Engineering, MIT

Dean Martin H. Bloom, College of Engineering, Polytechnic Institute of Brooklyn

Dean Harold A. Bolz, College of Engineering, Ohio State University

Dr. John F. Connors, Director, Professional Development, Martin Marietta Corporation

Dean Elmer C. Easton, College of Engineering, Rutgers—The State University

Dr. Dolph G. Ebeling, Consultant, Engineering Consulting and Education, General Electric Co.

Dr. Arthur G. Hansen, President, The Georgia Institute of Technology

Dr. J. W. Haun, Vice President and Director of Engineering, General Mills, Inc.

Mr. Hardy Liston, Jr., Chairman, Mechanical Engineering Department, School of Engineering, A & T College of North Carolina

Mr. M. R. Lohmann, Office of the Dean, College of Engineering, Oklahoma State University, Stillwater, Oklahoma 74074

Dean James M. McKelvey, School of Engineering and Applied Science, Washington University

Dean W. W. Mullins, Carnegie Institute of Technology, Carnegie-Mellon University

Dean C. H. Norris, College of Engineering, University of Washington

Dean N. J. Palladino, College of Engineering, Pennsylvania State University

Mr. Frank G. Rizzardi, Manager, Technical Training, Systems Group of TRW, Inc.

Dean Chauncy Starr, College of Engineering and Applied Science, University of California at Los Angeles

Dr. D. L. Thomsen, Jr., Director, Engineering Education, IBM Corporation

Dean Robert E. Uhrig, College of Engineering, University of Florida

Professional Engineering Career Development Series

APPLIED ENGINEERING STATISTICS FOR PRACTICING ENGINEERS

Lawrence Mann, Jr.

Mechanical, Aerospace & Industrial Engineering Department
Louisiana State University

BARNES & NOBLE, INC. NEW YORK

Publishers • Booksellers • Since 1873

L. C. Catalogue Card Number: 76-129578

SBN 389 00508 8

Distributed

In Canada
by the Ryerson Press, Toronto

In Australia and New Zealand
by Hicks, Smith & Sons Pty. Ltd.,
Sydney and Wellington

In the United Kingdom, Europe, and South Africa
by Chapman & Hall Ltd., London

Printed in the United States of America

Preface

This book results from the formalization of a set of notes prepared for in-plant seminars and short courses. In presenting it for use in these courses, the author has tried to make it understandable to practicing engineers.

The emphasis throughout is on practical problems encountered in technical decision-making situations. It is anticipated that the many examples, worked out in detail, will enable the reader to pursue the subject on his own. This work should be of particular interest to the engineer who graduated five to fifteen years ago and who encountered little or no use of statistical methods in his curriculum. This book will also serve as a reference for the engineer who occasionally finds the need to reinforce his decision by using statistical methods.

The mathematical background expected of the reader is no more than would be expected of the average practicing engineer.

I am indebted to the Literary Executor of the late Sir Ronald A. Fisher, F.R.S., to Dr. Frank Yates, F.R.S., and to Oliver & Boyd, Ltd., Edinburgh, for permission to reprint Table IV from their book *Statistical Tables for Biological, Agricultural and Medical Research* and the Table of t values from the work *Statistical Methods for Research Workers* by the same author and publisher. I am also indebted to The MacMillan Company for their permission to use the Table of z values from their publication entitled *Introduction to Probability and Statistics*, 2nd edition, by B. W. Lingdren and G. W. MacElrath.

Contents

1
Probability

"TIME AND CHANCE HAPPENETH."
Ecclesiastes (Chapter 9, Verse 11)

Since we are not able to control chance, we do the best we can; i.e., we try to quantify or predict chance. When we reflect upon an event that has not yet occurred (assuming the outcome is beyond our influence), we make use of the laws of chance outcome. Quoting from the Life Science Library volume on mathematics, "...probability is a percentage: the frequency with which one phenomenon takes place in relation to possible alternatives. When combined, the probabilities of particular events can be used to evaluate the chances of chains of events. To deal with such combinations, certain basic rules have been formulated; it is these rules which have come to be known as the laws of chance."

The engineer's work is replete with examples having probabilistic bases. He orders a certain item of equipment; the supplier gives him a delivery date of 90 days. What is the probability that he will meet his delivery date? He designs a machine to operate over a life span of 5000 cycles. What is the probability that the machine will last 5000 cycles? The engineer designs a heat exchanger to transfer heat at a certain rate under some given conditions. What is the probability that it will meet specifications? Since very few engineers agree that these three events will occur with certainty, the next best thing is to *quantify* the probabilities. Once quantified, the probabilities can then become inputs into our decision-making process.

1.1 CLASSIC DEFINITION OF PROBABILITY

Assume an event, E, can happen r ways out of a total of n possible ways all of which are equally likely to occur. In this case the probability of occurrence of the event is

$$p = P(E) = r/n \tag{1.1}$$

The probability of the event not occurring is

$$q = 1 - p = 1 - r/n = \frac{n - r}{n}$$

Note that $p + q = 1$.

Example: Let E be the event that the number 5 turns up in one toss of a die. There are six ways that the die can fall. If E can occur in only one of these ways, then $p = P(E) = \frac{1}{6}$, and the probability of not getting a 5 is $q = 1 - p = 1 - \frac{1}{6} = \frac{5}{6}$.

The probability of any event, or string of events, must fall between 0 and 1, inclusive; i.e., $0 \leq P(E) \leq 1$. Certainty of occurrence is 1; certainty of nonoccurrence is 0.

1.2 EMPIRICAL PROBABILITY

The classic definition of probability depends on the "equally likely" concept. In engineering areas this concept often presents problems. Therefore the statistical or empirical definition of probability is advocated by many. The empirical definition equates the relative frequency distribution to the probability distribution. It is in this area of investigation that most engineers are working today, that is, the relating of physical occurrences to recognized mathematical distributions and the creation of algorithms and models to use this information in design and decision-making processes.

Example: If a roulette wheel is spun 3800 times, the number 7 should appear in 100 of the events, i.e., with a probability of 0.0263 (36 numbers plus the 0 and the 00). If the wheel is actually spun and the 7 appears 106 times, then the empirical probability is 0.0279 versus 0.0263 for the classic case.

Conceptually, the idea of empirical probability is closely allied to the learning process. A child drops a toy balloon four times, but it bursts on the fifth try, so he learns that every once in a while the balloon will break when dropped. An engineer experiments with a new material and records the working stresses. These he uses in his design calculations but occasionally the material fails. He learns to rely on his knowledge if the failure is relatively rare.

Actually, it might be conjectured that for a given series of events, the empirical probability approaches some classic or ideal probability distribution as the number of events approach infinity. Since this is very expensive, and in some cases impossible, in engineering work, the conjecture cannot become an assumption, so we must make use of the empirical probability.

1.3 SOME PROBABILISTIC RELATIONSHIPS

If E_1 and E_2 are events and if one occurs and the other is in no way affected, then the events are termed *independent*. In this case the probability of both events is equal to the probabilities of each separate event.

$$P(E_1 \text{ and } E_2) = P(E_1)\,P(E_2) \tag{1.2}$$

Example: The probability of drawing a deuce, or any card for that matter, is $\frac{4}{52}$. The probability of drawing two deuces, each from a separate deck, is equal to $\left(\frac{4}{52}\right)\left(\frac{4}{52}\right) = \frac{1}{169} = 0.0059$.

At this point let us digress to define the notation terms which will be used from this point onward.

E_i = any of the n values $E_1, E_2, E_3, \ldots, E_n$. The letter i denotes any of the numbers $1, 2, 3, \ldots, n$ and is called the index or subscript.

$\prod_{i=1}^{r} P(E_i)$ = the *product* of the separate probabilities starting with subscript 1 and continuing through the rth.

$\sum_{i=1}^{r} P(E_i)$ = the *sum* of the separate probabilities starting with the subscript 1 and continuing through the rth.

If many events occur and the separate probabilities of occurrence do not affect one another, then the chance of them all happening is equal to the product of all the separate probabilities.

$$P(E_1 E_2 \cdots E_n) = \prod_{i=1}^{n} P(E_i) \tag{1.3}$$

This is known as the multiplicative law of probability.

Two or more events are *mutually exclusive* if the occurrence of any one of them excludes the occurrence of the remaining ones.

If E_1 and E_2 are independent, mutually exclusive events, then the probability of E_1 or E_2 is the sum of the individual probabilities.

$$P(E_1 \text{ or } E_2) = P(E_1) + P(E_2) \tag{1.4}$$

Example: The probability of drawing either a deuce or a trey in one draw from a single deck of cards is $\frac{4}{52} + \frac{4}{52} = \frac{8}{52} = 0.154$.

If more than one event is involved, and if those events are independent and mutually exclusive, then the probability is the sum of the individual probabilities.

$$P(E_1 \text{ or } E_2 \text{ or } E_3 \text{ or } \cdots E_n) = \sum_{i=1}^{m} P(E_i) \tag{1.5}$$

This is known as the additive law of probability.

If E_1 and E_2 are independent, but not mutually exclusive, then the probability of at least one of the events is the sum of the individual probabilities minus the probability of both events.

$$P(E_1 \text{ and/or } E_2) = P(E_1) + P(E_2) - P(E_1 E_2) \tag{1.6}$$

Example: The probability of drawing at least one face card in two draws, each from a separate deck, is $\frac{12}{52} + \frac{12}{52} - \frac{9}{169} = 0.409$.

1.4 CONDITIONAL PROBABILITY

If an event E_2 is dependent on the prior occurrence of event E_1, then the probability of E_2 is called a conditional probability. E_2 occurs on the condition that E_1 has occurred and is designated $P(E_2 \mid E_1)$. The probability of a compound event, E_1 and E_2, is equal to the probability of one event multiplied by the conditional probability of the remaining event conditional on the first event having happened.

$$\begin{aligned} P(E_1 E_2) &= P(E_1)\, P(E_2 \mid E_1) \\ &= P(E_2)\, P(E_1 \mid E_2) \end{aligned}$$

Example: Assuming one deck of cards, what is the probability of drawing two aces on two draws? We do not replace the first ace after having drawn it (known as "without replacement").

$$\left(\frac{4}{52}\right)\left(\frac{3}{51}\right) = \frac{12}{2652} = 0.0045$$

Example: A box contains four white balls and three red balls. E_1 is the event that the first ball drawn is red and E_2 is the event that the second ball drawn is red, without replacement. (Note that E_1 and E_2 are dependent events.)

$$P(E_1) = \frac{3}{4+3} = \frac{3}{7} = \text{probability that first ball drawn is red}$$

$$P(E_2 \mid E_1) = \frac{2}{4+2} = \frac{2}{6} = \frac{1}{3} = \text{probability that second ball drawn is red given that the first ball drawn was red}$$

The probability that both balls drawn were red is

$$P(E_1 E_2) = \frac{3}{7} \cdot \frac{1}{3} = \frac{3}{21} = \frac{1}{7}.$$

1.5 BAYES' THEOREM

Bayes' theorem is a special case of conditional probability. It is used to establish the probability that an event, previously occurred, might have happened in a particular manner. If an event A is dependent on one of a number of mutually exclusive events $E_1, E_2, \ldots, E_n$, and if A has already happened, then the probability that event E_i has also happened is

$$P(E_i \mid A) = \frac{P(E_i)\, P(A \mid E_i)}{\Sigma\, P(E_i)\, P(A \mid E_i)}. \qquad (1.7)$$

Example: We have quality-control information on control valves received by the plant from two suppliers, E_1 and E_2. We receive three times as many valves from E_1 as E_2. Therefore the probability that any valve came from E_1 is 0.75, and from E_2 is 0.25. From experience we know that 90% of the valves from E_1 meet the specifications and 85% of the valves from E_2 are satisfactory. We would like to know (event A in our nomenclature) the probability that any one valve is satisfactory.

$$\begin{aligned} P(E_1) &= 0.75 \\ P(E_2) &= 0.25 \\ P(A \mid E_1) &= 0.90 \\ P(A \mid E_2) &= 0.85 \end{aligned}$$

Therefore

$$\begin{aligned} P(A) &= P(E_1)\,P(A \mid E_1) + P(E_2)\,P(A \mid E_2) \\ &= 0.75(0.90) + 0.25(0.85) \\ &= 0.888 \end{aligned}$$

Now, enlarging on this problem, assume a third supplier E_3 so that the input data into the problem change as follows:

$$\begin{aligned} P(E_1) &= 0.70 & \quad P(A \mid E_1) &= 0.90 \\ P(E_2) &= 0.20 & \quad P(A \mid E_2) &= 0.80 \\ P(E_3) &= 0.10 & \quad P(A \mid E_3) &= 0.85 \end{aligned}$$

Our problem now is this: When we encounter a valve that operates properly, what is the probability that it came from E_3?

Bayes' expression, then, becomes for this particular case

$$P(E_3 \mid A) = \frac{P(E_3)\,P(A \mid E_3)}{\sum_{i=1}^{3} P(E_i)\,P(A \mid E_i)}$$

where the denominator is

$$\begin{aligned} 0.70(0.90) &= 0.630 \\ 0.20(0.80) &= 0.160 \\ 0.10(0.85) &= \underline{0.085} \\ & 0.875 = \Sigma\, P(E_i)\,P(A \mid E_i) \end{aligned}$$

Therefore

$$P(E_3 \mid A) = \frac{0.10(0.85)}{0.875} = 0.0972$$

1.6 PERMUTATIONS

The number of permutations of n items which are different is the number of different arrangements in which the items can be placed. It is possible to take all n items each time or r items at a time, where there are less r than n. The identity *and order* in which the items are arranged is of importance.

The number of permutations of n objects taken r at a time is denoted by nPr, $P(n, r)$, P_r^n or Pn, r.

In the generalized expression there are r spaces to fill with n items; therefore

The first space can be filled in n ways,
The second space can be filled in $n - 1$ ways,

$\vdots$

The rth space can be filled in $n - r + 1$ ways.

From this we get $nPr = n(n - 1)(n - 2)\cdots(n - r + 1)$, which (without algebraic proof) reduces to

$$\frac{n!}{(n - r)!} \tag{1.8}$$

and as a special case $nPn = n!$, remembering that, by definition, $0! = 1$.

Example: How many three-digit numbers can be formed using the numbers 0, 1, 2, 3, 4, and 5 if each number can be repeated?

The first digit can be filled in six ways as can the second and third digits. Therefore

$$6 \times 6 \times 6 = 216 \text{ different numbers}$$

Example: How many three-digit numbers can be formed

using the numbers 0, 1, 2, 3, 4, and 5 if the numbers *cannot* be repeated?

$$nPr = \frac{n!}{(n-r)!} = \frac{6!}{3!} = 6 \cdot 5 \cdot 4 = 120 \text{ numbers}$$

1.7 STIRLING'S APPROXIMATION TO n!

When n is large, $n!$ is difficult to evaluate. When this situation is encountered, use can be made of Stirling's approximation, which is

$$n! \approx \sqrt{2\pi n}\, n^n e^{-n}$$

1.8 COMBINATIONS

The number of combinations of n different items is the number of different groups of r items each, *disregarding* order and arrangement. The number of combinations of n objects taken r at a time is denoted by nCr, $C(n, r)$, $C_{n,r}$ or $\binom{n}{r}$.

The generalized expression for r items that can be arranged in $r!$ ways is

The first space can be filled in r ways,
The second space can be filled in $r - 1$ ways,

⋮

The rth space can be filled in 1 way.

From this it can be seen that r spaces can be filled in $r(r-1)\cdots 1 = r!$ ways, which (without algebraic proof) reduces to

$$\frac{n!}{(n-r)!r!} \tag{1.9}$$

Example: How many separate 3-man crews can be formed from a group of 10 men? Here $n = 10$ and $r = 3$; therefore

$$\binom{n}{r} = \frac{n!}{(n-r!)r!} = \frac{10!}{(10-3)!3!} = \frac{3{,}628{,}800}{5040(6)} = 120$$

Note that probability is the number of desirable outcomes divided by all possible ways that the event can occur. Therefore

$$P(E) = \frac{\binom{n_1}{r_1}}{\binom{n_2}{r_2}} \tag{1.10}$$

So, if we wanted to know the probability of drawing a bridge hand with all four aces, we would calculate it in this manner.

For the denominator (that is, how many possible 13-card hands are there in a 52-card deck) we have $\binom{52}{13}$, which equates to 635,013,559,600.

For the numerator, there is only one way to get all four aces and the remainder of the hand can be filled with any 9 cards of the 48 remaining. So the numerator will be $1\binom{48}{9}$, which is 1,677,106,600. Therefore

$$P(4\text{ aces}) = \frac{1{,}677{,}106{,}600}{635{,}013{,}599{,}600} = 0.0026 \qquad \text{or} \qquad \text{about 1 in 379}$$

The above is not so difficult to calculate as it seems. Remembering Eq. (1.10), we have

$$P(4\text{ aces}) = \frac{\binom{48}{9}}{\binom{52}{13}} = \frac{48!}{39!\,9!}\,\frac{39!\,13!}{52!} = \frac{13 \cdot 12 \cdot 11 \cdot 10}{52 \cdot 51 \cdot 50 \cdot 49}$$

$$= \frac{17{,}160}{6{,}497{,}400} = 0.0026$$

These large combinatorial factorials can also be solved by logarithms. In the case of the 5-card poker hand, we have

$$\log\binom{52}{5} = \log 52! - \log 5! - \log 47!$$

$$\log 52! = 67.90665$$

$$\log 5! = 2.07918$$

$$\log 47! = 59.41267$$

$$
\begin{aligned}
\therefore \log \binom{52}{5} &= \log 52! - \log 5! - \log 47! \\
&= 67.90665 - 2.07918 - 59.41267 \\
&= 6.41480 \\
&= 2.599 \times 10^6 = 2{,}599{,}000
\end{aligned}
$$

or, to be more accurate, 2,598,960.

Tables of $n!$ and $\log n!$ are available in handbooks.

2
Statistical Parameters

Prior to the major task of acquiring an understanding of distributions, it is necessary to consider the mathematical characteristics by which we identify distributions. These mathematical characteristics are called *parameters*. In short, a parameter is a quantity that describes a statistical population.

The minimum information necessary to describe any distribution is a measure of central tendency and a measure of variability. Before investigating the parameters themselves, consider the following situation to illustrate the need for both above mentioned parameters to describe adequately a distribution.

A series of data points is made on two different chemical processes as shown in Table 2.1. The average for process A is 56.0/10 or 5.6, and the average for process B is, similarly, 56.0/10 or 5.6. Since the two statistical populations have the same average, does this mean they are similar?

TABLE 2.1

Process A	Process B
5.8	3.6
6.1	4.5
4.9	7.1
5.2	4.5
6.0	7.3
5.5	6.8
5.4	4.3
5.7	6.4
6.0	4.5
5.4	7.0
56.0	56.0

A glance at the figures indicates that the data from process *B* are more variable than *A*. We have an inherent feeling (borne out by mathematical proof) that process *A* is better and our intuition is based on the variability aspect. But more about variability later.

2.1 MEASURES OF CENTRAL TENDENCY

The average or mean (sometimes called the *arithmetic mean*) is the most common measure of central tendency. It is also called the *expected value*.

The arithmetic mean of a set of *n* numbers, $x_1, x_2, x_3, \ldots, x_n$, is abbreviated using the symbol $\bar{x}$ and is

$$\bar{x} = \frac{x_1 + x_2 + \cdots + x_n}{n} = \frac{\sum_{j=1}^{n} x_j}{n} = \frac{\Sigma x}{n} \qquad (2.1)$$

Example: The arithmetic mean of the numbers 12, 18, 15, 14, and 17 is

$$\bar{x} = \frac{11 + 18 + 15 + 14 + 17}{5} = \frac{75}{5} = 15$$

Frequently, engineering data are in such a form that it is tedious to work with in its raw state. In this case it is necessary to group the data into class intervals to obtain a workable frequency distribution.

Consider the data in Table 2.2.

TABLE 2.2

Outside Diameter of 50 Cylinders, in 0.0001 "over 3.000"				
14	15	11	15	20
15	18	13	17	18
13	14	22	18	15
12	15	19	16	15
15	12	18	16	14
15	13	16	13	15
14	15	13	15	18
18	16	13	15	14
14	18	18	17	15
14	16	15	18	15

Grouping the data we get

11–12	111				3
13–14	~~1111~~	~~1111~~	111		13
15–16	~~1111~~	~~1111~~	~~1111~~	~~1111~~	20
17–18	~~1111~~	~~1111~~	1		11
19–20	11				2
21–22	1				1

For the data in Table 2.2, the first *class limits* are 11 and 12. *Class boundaries* denote the most extreme true values included in the class. Since the "11" means a measurement to the nearest ten-thousandth of an inch, any dimension actually lying between 10.50 and 11.50 ten-thousandths will be tallied as an 11. Therefore, the class boundaries of the first grouping are 10.5 and 12.5. *Midvalue* or *class mark* is the value which is halfway between the two boundaries of the class. Width of the class or *class interval* is the difference between two consecutive midvalues. Thus, in Table 2.2, the class interval is 2 "tenths," because this is the difference between the midvalues 11.5 and 13.5.

The expression for computing the arithmetic mean when the data are grouped is

$$\bar{x} = M + \left(\frac{\sum_{i=1}^{n'} f_i u_i}{n} \right) c \tag{2.2}$$

where $\bar{x}$ = arithmetic mean
M = midvalue of class in which mean is assumed
f_i = frequency of occurrence in ith class interval
u_i = indexing or weighting integers (see following example)
c = class interval size
n = total number of items in data (not the number of class intervals)
n' = total number of class intervals

Example: Grouping the data from Table 2.2, we get

Midvalue of Class Interval	Frequency f	Index u	fu
11.5	3	−2	−6
13.5	13	−1	−13

Midvalue of Class Interval	Frequency f	Index u	fu
15.5	20	0	0
17.5	11	1	11
19.5	2	2	4
21.5	1	3	3
			$\Sigma = -1$

Therefore $M = 15.5$, $\Sigma fu = -1$, $n = 50$, and $c = 2$, and so

$$\begin{aligned}\bar{x} &= 15.5 + \left(\frac{-1}{50}\right)2 \\ &= 15.5 + (-0.02)\,2 \\ &= 15.5 - 0.04 \\ &= 15.46\end{aligned}$$

Note that it makes no difference if the class interval where the mean lies is assumed in error; the value of Σfu will merely change, yielding the same answer.

There are other measures of central tendency and the next to be considered is the *median*. The median of a set of values is the middle value when the observations are arranged in either ascending or descending order. The median is sometimes defined as that value above and below which an equal number of values occur.

Example: Consider the integers 8, 9, 9, 7, 10, 7, 9, 9, 8, 10, 7.

Arranging the data in rank order, we have 7, 7, 7, 8, 8, ⑨, 9, 9, 9, 10, 10. It can be seen that the circled 9 has five data points below and above it. Therefore, it is the median value. If there are an even number of values, then the arithmetic mean of the two central most values becomes the median.

If the data are grouped, then the median can be obtained (by interpolation) from the following expression:

$$x_{\text{me}} = B + \left[\frac{(n/2) - (\Sigma f_i)_m}{f_m}\right] c \qquad (2.3)$$

where x_{me} = median

B = lower class boundary of the class in which it is assumed that the median lies

$(\Sigma f_i)_m$ = summation of frequencies of all classes lower than the class in which it is assumed that the median lies (exclusive)

f_m = frequency of the class in which it is believed that the median lies

c, n = as before

Example: Again, considering the data in Table 2.2 and substituting in (2.3) where

$$B = 14.5$$
$$(\Sigma f_i)_m = 13 + 3 = 16$$
$$c = 2$$
$$n = 50$$
$$f_m = 20$$
$$\therefore x_{\text{me}} = 14.5 + \left(\frac{(50/2) - 16}{20}\right)2 = 15.40$$

which compares rather favorably with the 15.46 value which was obtained for the mean.

Another measure of central tendency is the mode. The *mode* is defined as the most frequently appearing value. A distribution does not have to have a mode. Conversely, a distribution may have more than one mode. At times the term is used rather loosely and peaks of distributions are called modes even though the specific peak under consideration does not reach the magnitude of the true mode.

In the data used as an example to explain the median, 9 is the most frequently appearing value. Therefore 9 is the mode.

Similar to the mean and median, the mode can be calculated from grouped data, using the following model:

$$x_{\text{mo}} = B + \left(\frac{D_l}{D_l + D_u}\right)c \tag{2.4}$$

where x_{mo} = mode

D_l = difference between frequency of class interval in which it is assumed that the mode lies and frequency of the next lower class interval

D_u = difference between frequency of class interval in which it is assumed that the mode lies and frequency of the next higher class interval

B, c = as before

Example: Compute the mode for the grouped data shown in Table 2.2.

Substituting in model (2.4),

$$x_{mo} = 14.5 + \left(\frac{7}{7+9}\right)2 = 14.50 + 0.88 = 15.38$$

Again, this compares with 15.70 for the median and 15.46 for the mean.

Why is there a difference between these three measures of central tendency? This difference is due to the fact that the distribution is somewhat skewed. *Skewness* is the degree to which a distribution departs from symmetry. If a distribution leans to the right (the long tail being on the left), then it is said to be skewed to the left (negative skewness); if it leans to the left (the long tail being on the right), then it is said to be skewed to the right (positive skewness). For skewed distributions the mean usually lies on the same side of the mode as the longer tail. This enables one to obtain a measure of nonsymmetry by employing the measure of the difference between the mean and the mode. See Figure 2.1. But the abovementioned difference is not the numerical value of skewness.

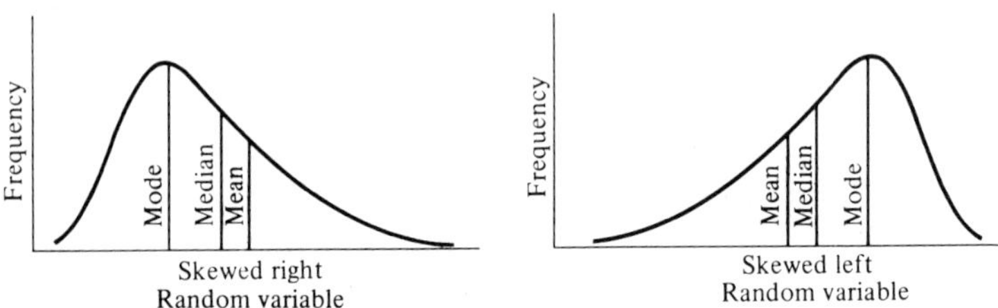

Fig. 2.1. Illustration of skewed distributions.

That model must wait until the discussion of the measures of variability has been completed. It is interesting to note, from Figure 2.1, that the mean passes through the centroid of the area, that the median divides the distribution into two halves, and that the mode is the value corresponding to the highest frequency.

There are other measures of central tendency such as the harmonic mean and the geometric mean, but they are not necessary to the subject under discussion.

2.2 MEASURES OF VARIABILITY

The second parameter necessary to the most basic description of any distribution is a measure of variability or dispersion. As in the measures of central tendency, there are a number of "yardsticks" by which variability can be measured.

Perhaps the most common and most widely used measure of variability is the range. The *range* is merely the difference between the highest and lowest value of any distribution. Since the range is dependent on only two values, its use is rather limited and it is efficient only for small samples.

Example: What is the range of the data shown in Table 2.2?

The highest value is 22 and the lowest value is 11. Therefore $R = 22 - 11 = 11$.

Another measure of dispersion is the standard deviation. The *standard deviation* is defined as the root mean square of the deviation from the mean and is defined as

$$s = \sqrt{\frac{\sum_{i=1}^{n} (x_i - \overline{x})^2}{n - 1}} \tag{2.5}$$

where s = standard deviation
x_i = any value in the population
$\overline{x}$ = mean of all the values
n = number of values in the population.

It is necessary here to digress to consider the aspect of statistical *degrees of freedom.* Note that the denominator under the radical in (2.5) is $n - 1$. Why not merely n? The degrees of freedom associated are the number of independent observations available for consideration. In calculating $\overline{x}$, all values of the sample were used. Thus, if all but one of the observations are known and $\overline{x}$ is known, the remaining observation can be obtained by difference. So, because 1 degree of freedom is used to obtain the mean, that observation is lost, and hence $n - 1$ becomes the denominator. Therefore the degrees of freedom can be defined, for purposes of this work, as the number of independent observations minus the number of population parameters employed or the number of constants calculated from the data.

While we are digressing it may be well here to emphasize

that the parameters with which we are working are *estimates* of the population. When we say that the mean is j and that the standard deviation is k, we actually should state that "the *estimate* of the mean is j and the *estimate* of the standard deviation is k." Since we are dealing with samples of the population, all information gained from that sample are merely estimates. Convention uses lowercase Latin letters to indicate sample parameters and lowercase Greek letters for population parameters. Thus m may indicate sample mean and μ population means; s may indicate sample standard deviation and σ population standard deviation; etc.

Returning to the explanation of the standard deviation, consider the following:

Example: Calculate the standard deviation of the following data: 6.2, 5.5, 6.4, 5.8, 4.9, 7.2.

Arrange the data in tabular order so as to accumulate the information needed to substitute in (2.5).

x_i	$\bar{x}$	$x_i - \bar{x}$	$(x_i - \bar{x})^2$
6.2	6.0	0.2	0.04
5.5	6.0	−0.5	0.25
6.4	6.0	0.4	0.16
5.8	6.0	−0.2	0.04
4.9	6.0	−1.1	1.21
7.2	6.0	1.2	1.44
6.⌊36.0			Σ = 3.14

$$\therefore \bar{x} = 6.0$$

$$\therefore s = \sqrt{\frac{3.14}{5}} = \sqrt{0.629} = 0.794$$

What is the physical meaning of an s of 0.794? This is a quantitative measure of the spread of the data. Spread, in a general sense, indicates lack of consistency. Usually, the more consistent the data, the more confidence we have that the sample truly represents the population. The standard deviation is measured in terms of the same units as the sample units. (We will later learn to use the term *random variable* as a general term for the scale under discussion.)

For grouped data, the expression for the standard deviation is

$$\therefore s = c\sqrt{\frac{\Sigma f_i (u_i)^2}{n} - \left(\frac{\Sigma f_i u_i}{n}\right)^2} \tag{2.6}$$

where s is the standard deviation and the other terms are as defined for model (2.2). [n is used here instead of $n - 1$ because no parameter is calculated to obtain s in model (2.6).]

Example: Using the data from Table 2.2 we get

Class Interval	f	u	fu	$f(u)^2$
11–12	3	−2	−6	12
13–14	13	−1	−13	13
15–16	20	0	0	0
17–18	11	1	11	11
19–20	2	2	4	8
21–22	1	3	3	9
			−1	53

Substituting in (2.6),

$$s = 2\sqrt{\frac{53}{50} - \left(\frac{-1}{50}\right)^2} = 2\sqrt{1.0596} = 2.06$$

Now that an understanding of the standard deviation has been acquired, it is possible to define the quantitative model for skewness (dimensionless) as

$$\text{Skewness} = \frac{\text{mean} - \text{mode}}{\text{standard deviation}} = \frac{\bar{x} - x_{\text{mo}}}{s}. \tag{2.7}$$

When the data are not grouped, it is sometimes very time consuming to use (2.5) to obtain the standard deviation. Where a desk calculator is employed, the following expressions should be used:

$$s = \sqrt{\frac{n\,\Sigma x_i^2}{n^2} - \left(\frac{\Sigma x_i}{n}\right)^2} = \frac{\sqrt{n\,\Sigma (x_i)^2 - (\Sigma x_i)^2}}{n}$$

The quantity $\Sigma (x - \bar{x})^2$ can be obtained more easily with a desk calculator if model (2.8) is used:

$$\Sigma (x - \bar{x})^2 = \Sigma x^2 - \bar{x}\,\Sigma x \tag{2.8}$$

At times it is more useful to eliminate the radical and define s^2 instead of s. The *variance* is defined as the square of the standard deviation.

As a preliminary toward the consideration of frequency distributions (Chapter 3), consider the following illustrative example using the data from Table 2.2.

Class Interval	f	Cumulative f	Percent	Cumulative Percent
11–12	3	3	6	6
13–14	13	16	26	32
15–16	20	36	40	72
17–18	11	47	22	94
19–20	2	49	4	98
21–22	1	50	2	100

Figure 2.2 shows the class intervals plotted against the cumulative percent.

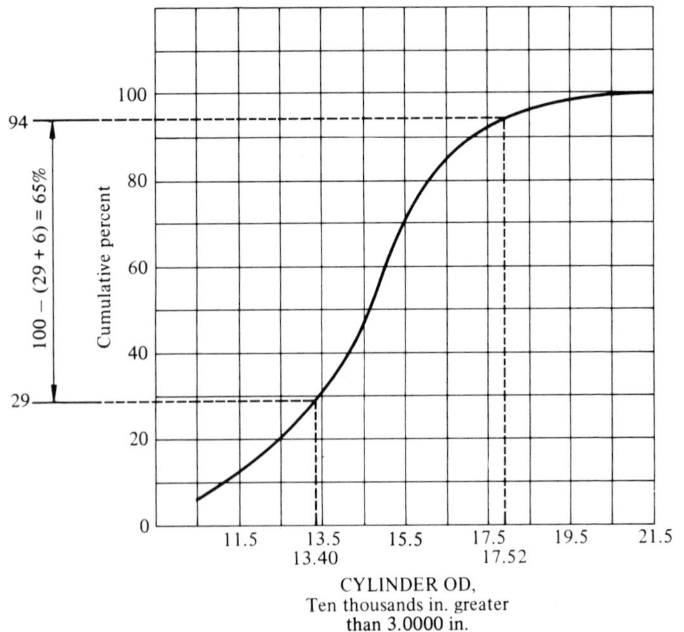

Fig. 2.2 Plot of cumulative percent curve of grouped data from Table 2.2.

It has previously been computed that the standard deviation for these data is 2.06 and that $\bar{x}$ is 15.46. Therefore

$$\bar{x} + 1s = 15.46 + 2.06 = 17.52$$
$$\bar{x} - 1s = 15.46 - 2.06 = 13.40$$

so if we locate these points on the cylinder OD scale and project them up to the cumulative curve and over to the cumulative percent scale, we can read 29 and 94% for the $-1s$ and $+1s$ points, respectively. This means that, for the sample under consideration, 65% (see Figure 2.2) of the area of the distribution is included between the $-1s$ and $+1s$ points. In a similar manner the $\pm 2s$ and $\pm 3s$ points can be calculated.

2.3 COEFFICIENT OF VARIATION

The units in which the standard deviation is expressed, as has been noted, are in the same units as are the class intervals. For many purposes, especially where unlike items are compared, a measure of variability, in percent terms, would be helpful. Such a measure exists and is termed the *coefficient of variation*, which is defined as

$$V = \frac{s}{\bar{x}} \times 100, \tag{2.9}$$

where V is the coefficient of variation and s and $\bar{x}$ are as defined previously. Note that the coefficient of variation is a dimensionless quantity.

Example: A manufacturer of electric lamps has two models of the same wattage bulb. Twenty samples of bulb A have a mean life of 1620 hr with a standard deviation of 300 hr. Twenty samples of bulb B have a mean life of 1940 hr with a standard deviation of 345 hr. Which lamp has the greatest relative variation?

$$\text{Coefficient of variation of } A = \frac{s}{\bar{x}} \times 100 = \frac{300}{1620} \times 100 = 18.5\%$$

$$\text{Coefficient of variation of } B = \frac{s}{\bar{x}} \times 100 = \frac{345}{1940} \times 100 = 17.7\%$$

Therefore lamp *A* has the greatest variability.

The coefficient of variation is useful in indicating the relative consistency of a number of data groupings. Often, consistency is a measure of "high quality" of data.

3
Distributions

"THOSE WHO CANNOT REMEMBER THE PAST ARE CONDEMNED TO REPEAT IT." *George Santayana*

All that has been covered up to this point has been preparatory to considering distributions. Most phenomena occur in patterns, and a major task of statistical methods is the recognition of this pattern (distribution) so that some prediction can be made about the future behavior of the process in question. The preface quotation by Santayana indicates that no prior knowledge (or memory) of the past (past being the distribution of collected data) prevents one from using experience to predict the future. The general concept of a frequency distribution is at the heart of statistical reasoning.

To think of a set of varying data points as coming from some distribution is fundamental to statistical processes. Whenever data collection takes place, a sampling process has occurred. As mentioned before, the fact that we are dealing with but a sample of some (sometimes infinitely) large distribution should constantly be kept in mind. If the sample is small, our ability to predict the actual population (large distribution) is remote. As the sample size becomes larger, we can predict with greater confidence. If we had all possible data, we could predict with certainty. It is the purpose of this chapter to explore different distributions and to learn how knowledge of these distributions can fit into our predictive and decision-making processes. For the sample to be truly representative of the population, it must be drawn without bias, which means that any data point in the population has an "equally likely" chance of being chosen for the sample.

A sample chosen in the manner indicated above is a random sample. *Randomness* indicates the complete lack of pattern or distribution. A *statistic* is a single value, calculated from sample data for which there is some assurance that it represents the parameter that it is supposed to estimate. In general a statistic obtained from a random sample is a *random variable.* The fact that the variable is termed "random" does not indicate that the frequency distribution of the variable is random. It does indicate that the order of arrival of the different possible values does elude predictability. The term *frequency distribution* was mentioned above. A *frequency distribution* is a plot with any random variable as the abscissa and the frequency of that random variable as the ordinate. Figure 3.1 shows the frequency distribution plot for the data given in Table 2.2. Whenever the term *distribution* is used, it is a brief way of indicating that one is referring to a frequency distribution plot.

In connection with random variables, one is usually interested in the percentage of the area between, say, *a* and *b* on the random variable scale. In this connotation, we have a probability plot. For instance, in Figure 3.1, if, say, 40% of the area lies to the left of 15.5, we can say, if the data are a satisfactory representation of the population, 40% of the cylinders that we will ever make on this process will fall between 9.5 and 15.5

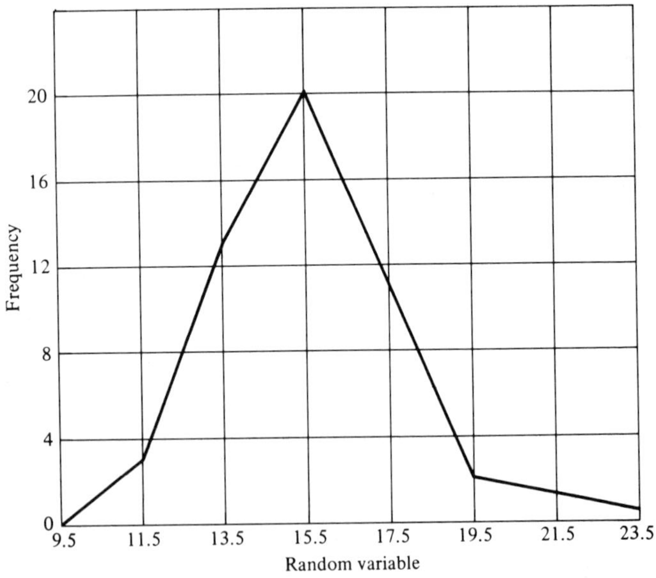

Fig. 3.1 Frequency distribution plot for data in Table 2.2

tenths over 3.0000 in. Thus, the frequency distribution plot becomes a probability distribution plot. If we view it as a probability distribution plot, there are some properties that should be defined. The area under the curve equals 1. Therefore portions of the area are always less than 1 and represent the probability of occurrence between any points, a and b, on the random variable scale. A word about notation. The term $f(r)$ or $f(x)$ indicates a function, usually over its entire range. The term $p(r)$ or $p(x)$ usually indicates some portion of the function in a probability sense. It is not unusual to find the terms used interchangeably because the distinction, in many cases, is a fine one. It will be seen later that x and r are used interchangeably, r being used largely where the gamma distribution is involved.

3.1 MATHEMATICAL EXPECTATION

The probability p that a piece of equipment will produce y pounds of a product is py. This concept is known as mathematical expectation. For example, if the door prize at a meeting is \$25 and there are 50 attendees, then the expectation of each attendee is $\frac{1}{50}(\$25) = \$0.50$.

Let x denote a discrete random variable which can assume the values $x_1, x_2, \ldots, x_n$, with the respective probabilities $p_1, p_2, \ldots, p_n$. The mathematical expectation of x, denoted by $E(x)$, is

$$E(x) = p_1 x_1 + p_2 x_2 + \cdots + p_n x_n = \sum_{i=1}^{n} p_i x_i = \Sigma px$$

If x is a continuous random variable (to be covered later) with a probability density function $f(x)$, then

$$E(x) = \int_{-\infty}^{\infty} px \, dx.$$

3.2 DISCRETE RANDOM VARIABLES

As we have seen, the distribution of a random variable is the definition of its mathematical behavior. In this section we will be dealing only with random variables which are discrete in nature, that is, variables which have a finite number of values. Since we are dealing with relatively few points on the random variable scale, we can assign a probability of occurrence at each point. We call this function the *probability mass function* (PMF) and at point i the PMF for a discrete random variable x is $p_x(x_i)$. Saying the same thing another way, the probability that the frequency of a random variable x is equal to x_i is $p_x(x_i)$. The PMF can be a plot such as shown in Fig. 3.1, a bar graph, or a histogram. Applying the property mentioned at the end of the previous section, we can state, for the discrete case, that there must be some value of the random variable associated with every point on the random variable scale (often 0); therefore

$$\sum_{i=1}^{\infty} p_x(x_i) = 1 \tag{3.1}$$

and it logically follows that $0 \leq p_x(x_i) \leq 1$ for all values of x_i. The very fact that we use the summation sign indicates that we are accumulating a finite number of points on the random variable scale.

By their own nature discrete variables are often of the discontinuous type, i.e., step functions. This plot is merely the tops of any histogram on a frequency distribution graph.

Binomial Distribution The first discrete random variable distribution which will be considered is the binomial. The binomial law, as it is sometimes called, is one of the most fundamental laws of probability. It is sometimes called the Bernoulli distribution or theorem after James Bernoulli (1654–1705), who was a member of a remarkable Swiss family which produced eight mathematicians of note in three generations. Johann Bernoulli did much to spread calculus in Europe.

The binomial expansion, familiar to all engineers, is $(p + q)^n$. If we expand this we get

$$(p + q)^n = p^n + nqp^{n-1} + \frac{n(n-1)}{2!} q^2 p^{n-2} + \frac{n(n-1)(n-2)}{3!} q^3 p^{n-3} + \cdots + q^n \tag{3.2}$$

It may be remembered from Chapter 1 that the coefficients in model (3.2) correspond to the combinatorial expression for n items taken 1, 2, 3, etc., at a time, respectively. Therefore rewriting (3.2) in combinatorial terms, we get

$$(p + q)^n = \binom{n}{0} q^0 p^{n-0} + \binom{n}{1} qp^{n-1} + \binom{n}{2} q^2 p^{n-2} + \cdots + \binom{n}{n} q^n p^{n-n}$$

and collecting terms, we finally get

$$(p + q)^n = \sum_{x=0}^{n} \binom{n}{x} q^x p^{n-x} \tag{3.3}$$

Since (3.3) accumulates all terms from $x = 0$ to n, the probability of occurrence of any individual value of x is

$$p(x) = \binom{n}{x} p^x q^{n-x} \tag{3.4}$$

where $p(x)$ = probability of occurrence of any value of x
n = total number of items in the sample
$x = 0, 1, 2, \ldots, n$
p = probability that the event sought will occur (sometimes called a "success")
$q = 1 - p$

Model (3.4) is the general expression for the binomial distribution. Note the variables necessary to plot the binomial distribution. They are

n samples
p probability of a certain event
x number of times event is encountered

Example: What is the probability of getting exactly two tails in six tosses of a coin?

p and q are mutually exclusive events such that

$$q = 1 - p \tag{3.5}$$

Although relationship (3.5) is often helpful, it is not necessary here since we already know p (the probability of getting heads) is $\frac{1}{2}$ and q (the probability of getting tails) is $\frac{1}{2}$. Therefore, recalling model (3.4), we get

$$p(x) = \binom{n}{x} p^x q^{n-x}$$

where

$$x = 2, \quad n = 6, \quad p = \tfrac{1}{2}, \quad q = \tfrac{1}{2}$$

$$\therefore p(x) = \frac{n!}{x!(n-x)!} p^x q^{n-x} = \frac{6!}{2!4!} \left(\frac{1}{2}\right)^2 \left(\frac{1}{2}\right)^2 = \frac{15}{64}$$

The binomial has an important role in quality-control work. Assume that we have an industrial process such that, from historical data, we know that we can expect 2% defectives in the process. If we test a sample and get 3% defectives, does this mean that we have encountered a *different* population? To make this decision we must know the probability of encountering 3% defectives when we are sampling from an overall 2% defective population. Actually we would like to know the probability of getting 0, 1, 2, etc., defectives. We will use this situation to illustrate further the binomial distribution as well as some others that we will encounter.

In the above-described problem, if we work with percentages, if $p = 0.02$, then $n = 100$, and $q = 1 - p = 1 - 0.02 = 0.98$, and $x = 0, 1, 2, \ldots, n$. Although it is theoretically possible to compute $p(x)$ for all possible x's, we will stop when $p(x)$ gets sufficiently small to be neglected. Therefore substituting in (3.4), for $x = 0$,

$$p(0) = \frac{100!}{0!\,100!} (0.02)^0 (0.98)^{100}$$

Remembering that $0! = 1$ and any number to the zero power equal 1, the above expression simplifies to

$$p(0) = 0.98^{100}$$

We will pause here to review the logarithmic manipulation to equate 0.98^{100}. A table of logarithms will be found as Appendix 1.

$$\log 0.98 = 9.9912261 - 10$$

so

$$100(9.9912261 - 10) = 999.12261 - 1000 = 9.12261 - 10$$

and from the log tables the antilog of 12261 is 1326 to the 10^{-1} power or 0.1326. Therefore

$$p(0) = 0.98^{100} = 0.1326$$

and so

$$p(1) = \frac{100!}{1!99!}(0.02)^1(0.98)^{99}$$

To evaluate the $p(1)$ expression, we use another short-cut method which prevents us from having to evaluate the logarithm for each iteration. We do this by multiplying the $p(1)$ expression by 0.98/0.98, which yields

$$p(1) = \frac{100(0.02)^1(0.98)^{100}}{0.98} = \frac{100}{49}[p(0)] = 0.2706$$

Using the same logic for $x = 2$, we get

$$p(2) = \frac{100(99)}{2(1)}(0.02)^2\frac{p(1)}{0.98} = \frac{99}{2}\frac{0.02}{0.98}[100(0.02)(0.98)^{99}]$$

Since the value inside the bracket is equal to $p(1)$, we have

$$p(2) = \frac{99}{2}\frac{0.02}{0.98}p(1) = \frac{99}{98}p(1) = 0.2734$$

For $x = 3$,

$$p(3) = \frac{100(99)98}{(3)(2)}(0.02)^3\frac{p(2)}{0.98}$$
$$= \frac{98}{3}\frac{0.02}{0.98}p(2) = 0.1823$$

For $x = 4$,

$$p(4) = \frac{97}{4}\frac{0.02}{0.98}p(3) = 0.0902$$

And so on. Note the symmetry of the expressions. Table 3.1 summarizes the results of this problem. If we plot the probability mass function depicted in Table 3.1, we get the curve shown in Figure 3.2, which is the typical configuration of the binomial distribution.

TABLE 3.1

Number of Defects in Sample, x	$p(x)$
0	0.1326
1	0.2706
2	0.2734
3	0.1823
4	0.0902
5	0.0354
6	0.0114
7	0.0031
8	0.0007
9	0.0002
	~1.0000

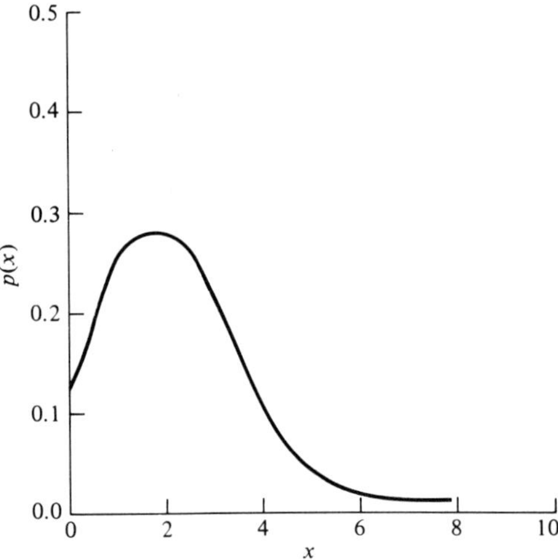

Fig. 3.2 Plot of binomial distribution from data in Table 3.1.

The binomial distribution is only theoretically correct when the sample size approaches infinity. Some properties of the binomial distribution are the

$$\text{Mean} = np$$
$$\text{Variance} = npq$$
$$\text{Skewness} = (q - p)/\sqrt{npq}$$

Poisson Distribution The Poisson distribution, so called because S. D. Poisson first recognized it in 1837, is sometimes called the "law of small numbers." It is an approximation to the binomial. In the binomial distribution, an event is called "rare" if n is large as p is small. Actually, if $n > 40$ while $np < 5$, an event is termed "rare." Within this definition of rarity, the Poisson distribution closely approximates the binomial. That is, if we let $n \rightarrow \infty$ and $p \rightarrow 0$ in such a way that np remains fixed at a value of the expected mean value, we obtain, from the binomial distribution, the Poisson distribution. Poisson distribution also represents independent events occurring at a constant rate (platoons of traffic, customers in groups, bunched defects, etc.).

The Poisson distribution is defined as

$$p(x) = \frac{\lambda^x e^{-\lambda}}{x!} \tag{3.6}$$

where $\lambda = np$ and the remainder of the terms are as defined previously. Note that the variables necessary to plot the Poisson distribution are merely λ, which is the expected mean value of the distribution in question. Merely knowing this, we can generate the distribution as contrasted with the binomial, where we must know n and p as definite values rather than as an expected mean defined by the product of n and p. Appendix 2 lists some values of e^x and e^{-x}. The usefulness of the table can be extended if it is remembered that, for example,

$$e^{-6.5} = (e^{-4.50})(e^{-2.00}) = (0.0111)(0.1353) = 0.0015$$

Example: Ten percent of the items of a certain process are defective. Determine the probability that, in a sample of 100 items, 2% will be defective.

$$p = 0.1, \qquad x = 2, \qquad \text{and} \qquad n = 100; \qquad \therefore \lambda = 10(0.1) = 1$$

and

$$p(2) = \frac{\lambda^x e^{-2}}{x!} = \frac{1^2 e^{-1}}{2!} = 0.184$$

Let us consider the same problem that we used to generate the binomial distribution but apply it to the Poisson. Remember that we had an industrial process where we expected 2% defectives (assuming a sample size of 100). Therefore

$$\lambda = np = 100(0.02) = 2$$

and so substituting in (3.6) for, first, $x = 0$,

$$p(0) = \frac{\lambda^x e^{-\lambda}}{x!} = \frac{2^0 e^{-2}}{0!}$$

and rearranging

$$p(0) = \frac{2^0}{0!} e^{-2}$$

remembering $0! = 1$ and $2^0 = 1$,

$$p(0) = e^{-2} = 0.1353$$
$$p(1) = \frac{2^1}{1!} e^{-2} = 0.2707$$
$$p(2) = \frac{2^2}{2!} e^{-2} = 0.2707$$

Note that the fact that $p(1) = p(2)$ is a coincidence and that

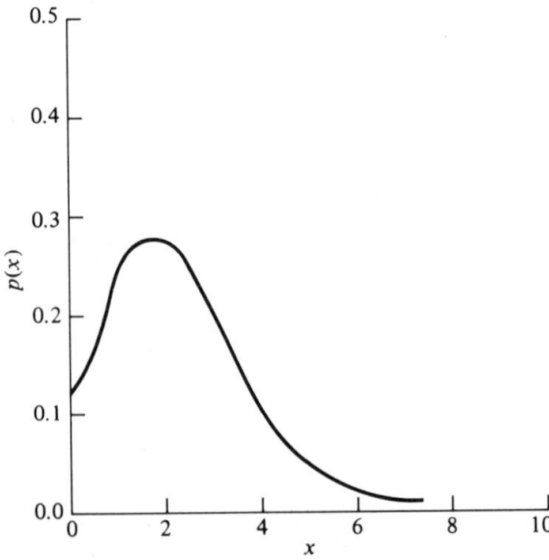

Fig. 3.3 Plot of Poisson distribution from data in Table 3.2.

no inference should be drawn from this example. Obviously $p(0) = e^{-\lambda}$ for all $x = 0$ cases of the Poisson.

$$p(3) = \frac{2^3}{3!} e^{-2} = 0.1804$$

$$p(4) = \frac{2^4}{4!} e^{-2} = 0.0902$$

Again, note the symmetry of the iterations. Table 3.2 summarizes the results of this problem. Figure 3.3, the plot of the distribution in Table 3.2, is typical of the Poisson distribution curve configuration.

TABLE 3.2

Number of Defects in Sample, x	$p(x)$
0	0.1353
1	0.2707
2	0.2707
3	0.1804
4	0.0902
5	0.0361
6	0.0120
7	0.0034
8	0.0009
9	0.0002
	~1.0000

Some properties of the Poisson distribution are that the

$$\text{Mean} = \lambda$$
$$\text{Variance} = \lambda$$
$$\text{Skewness} = 1/\sqrt{\lambda}$$

Hypergeometric Distribution A third distribution, closely related to the binomial, is the hypergeometric. This distribution applies (1) when both the population and sample size are large and when neither the number of defectives nor the population minus the number of defectives is very small and (2) to problems involving sampling from a small production lot. The expression for the hypergeometric distribution is

$$p(x) = \frac{\binom{N-d}{n-x}\binom{d}{x}}{\binom{N}{n}} \tag{3.7}$$

where N = population size
n = sample size
d = number defectives found in lot
x = probability of exactly x defectives

Note that we do not have a constant p here. Calculation of terms of (3.7) is quite tedious, so only a sample calculation will be given. Assume the following process:

$$N = 700$$
$$d = 14$$
$$n = 100$$
$$d/N = 14/700 = 0.02$$

Substituting in (3.7), we get, for $x = 0$,

$$p(0) = \frac{\binom{N-d}{n-x}\binom{d}{x}}{\binom{N}{n}} = \frac{\binom{686}{100}\binom{14}{0}}{\binom{700}{100}} = \frac{686!\,14!\,100!\,600!}{100!\,586!\,0!\,14!\,700!}$$

Canceling, we get

$$p(0) = \frac{686!\,600!}{586!\,700!}$$

and making use of any table of logarithms of factorials, we get

log 686! =	1649.6096	
log 600! =	1408.1023	
	3057.7119	3057.7119
log 586! =	1369.2745	
log 700! =	1689.3842	
	3058.6587	−3058.6587
		9.0532 − 10

The antilog of 9.0532 − 10 = 0.113.

$$\therefore p(0) = 0.113$$

This process can be continued until the distribution is generated similarly to that performed for binomial and Poisson distributions.

Some properties of the hypergeometric distribution are that the

$$\text{Mean} = nd/N$$

$$\text{Variance} = \frac{nd(N - d)(N - n)}{N^2(N - 1)}$$

Before going on, let us consider the three discrete distributions (binomial, Poisson, and hypergeometric) in summary.

For n at least 10 and p not over 0.1, the Poisson begins to be useful as an approximation to the binomial, and when n is 20 or more and p is 0.05 or less, the Poisson gives a very satisfactory approximation to the binomial. The Poisson is the distribution to use in situations where we are concerned with the number of defects on an item or sample of a product if the maximum possible number of such defects is much greater than the average number of them.

In general, we can use the binomial to approximate the hypergeometric if $n < 0.1N$ and use the Poisson if $p < 0.05$ and $n < 20$.

To illustrate further the close relationship of the three distributions, we will plot, on one set of coordinates, the curves obtained for the situation described on page 28 and previously used to illustrate the configurations of the binomial and the Poisson distributions. This plot is shown in Figure 3.4.

Multinomial Distributions The binomial distribution when generalized becomes the multinomial distribution. The name comes from the fact that the probabilities are the terms in the expansion of $(p_1 + \cdots + p_k)^n$. For $k = 2$, we have the binomial case. The expression for the multinomial distribution is

$$p(x_1, x_2, \ldots, x_p) = \frac{n!}{x_1!\,x_2!\cdots x_k!}\,p_1^{x_1}p_2^{x_2}\cdots p_k^{x_k} \tag{3.8}$$

where $x_i = 0, 1, \ldots, n$ such that $\sum_{i=1}^{k} x_i = n$. What we are really saying here is that we are considering the situation where we have n independent trials, each trial permitting k mutually

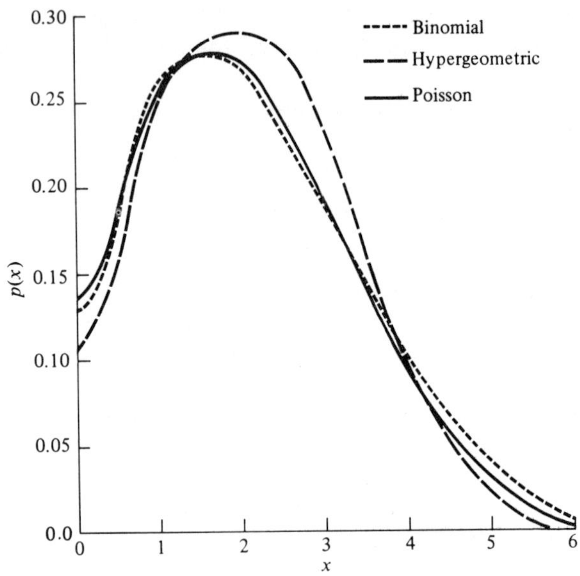

Fig. 3.4 Frequency plot comparing the binomial, hypergeometric, and Poisson distributions.

exclusive outcomes whose respective probabilities are $p_1, p_2, \ldots, p_k$, where $\sum_{i=1}^{k} p_i = 1$.

Example: The probability that a diode will fail in fewer than 400 hr is 0.40; the probability that it will fail in fewer than 700 hr but more than 400 hr is 0.3; and the probability that it will last more than 700 hr is 0.3. Find the probability that, among 12 such diodes, 4 will fail in fewer than 400 hr, 5 will fail in fewer than 700 but more than 400 hr, while 3 will last more than 700 hr.

Here we have $x_1 = 4$, $x_2 = 5$, $x_3 = 3$, $n = 12$, $p_1 = 0.4$, $p_2 = 0.3$, and $p_3 = 0.3$. Therefore, substituting in (3.8),

$$\begin{aligned} p(4, 5, 3) &= \frac{12!}{4!\,5!\,3!}(0.4)^4(0.3)^5(0.3)^3 \\ &= 27{,}720(0.0256)(0.0024)(0.0270) \\ &= 0.046 \end{aligned}$$

3.3 JOINT DISTRIBUTIONS OF DISCRETE VARIABLES

Thus far we have considered distributions with only one variate, commonly known as univariate distributions. We should be aware of the existence of bivariate and multivariate distributions. Many engineering problems possess a number of characteristics (variables) representing conditions of the problem. Once the student grasps the concepts of treating these types of problems, the transition to multidimensional applications is relatively easy.

Therefore, we will illustrate the concepts using a bivariate case. A product consists of two feed stocks produced in parallel processes each of which is graded 1, 2, or 3—the higher the number the better the product. Process *A* has the probability of 0.2 of producing quality 1, 0.4 of producing quality 2, and 0.4 of producing 3. Likewise, process *B*'s corresponding probabilities are 0.2, 0.3, and 0.5. The problem is to compute the joint distribution of the various quality mixes of the two processes assuming independent performance.

Let x_A indicate any possible quality of process *A* and let $p_A(x_A)$ be the probability of obtaining that quality; further, let x_B indicate any possible quality of process *B* and let $p_B(x_B)$ be the probability of obtaining that quality. Now, considering the bivariate aspect of the problem, let $p(x_A, x_B)$ be the probability of the compound event that process *A* has quality x_1 while concurrently process *B* has quality x_2. Remembering (1.3) from Chapter 1 and recalling the independence of the two events in question, we get

$$p(x_1, x_2) = p_A(x_1)\,p_B(x_2)$$

The probability that both processes produce quality-1 material is

$$p(1, 1) = p_A(1)\,p_B(1) = (0.2)(0.2) = 0.04$$

The probability that process *A* produces quality 1 and process *B* produces quality 2 is

$$p(1,2) = p_A(1)p_B(2) = (0.2)(0.3) = 0.06$$

If this process is repeated for all possible combinations, i.e., the complete bivariate distribution, the results are as tabulated in Table 3.3.

TABLE 3.3

x_A/x_B	1	2	3	$\sum = p_A x_A$
1	0.04	0.06	0.10	0.20
2	0.08	0.12	0.20	0.40
3	0.08	0.12	0.20	0.40
$\sum = p_B x_B$	0.20	0.30	0.50	1.00

Assume that the quality of the finished product is the sum of the individual qualities. We now want to find the distribution of the finished goods quality. This joint quality can assume the values of 2, 3, 4, 5, or 6. The probability of any of these values, $p(j)$, is equal to the sum of the probabilities of the different ways in which the combined qualities can be obtained (remember the "dice" example). They are

$$\begin{aligned}
p(2) &= p(1,1) && = 0.04 \\
p(3) &= p(1,2) + p(2,1) = 0.06 + 0.08 && = 0.14 \\
p(4) &= p(1,3) + p(2,2) + p(3,1) = 0.10 + 0.12 + 0.08 && = 0.30 \\
p(5) &= p(2,3) + p(3,2) = 0.20 + 0.12 && = 0.32 \\
p(6) &= p(3,3) && = \underline{0.20} \\
& && 1.00
\end{aligned}$$

Naturally, the total of all possible combinations must always equal 1.00. This process can be duplicated only if the variates are independent.

Other discrete distributions include the geometric, the Pascal, and the negative binomial.

3.4 CONTINUOUS RANDOM VARIABLES

In the previous section we considered distributions whose random variables assumed discrete, or integer, values. That is but one side of the coin. We must now consider distributions whose random variables can assume any or all of an infinite number of values. These distributions are called continuous statistical distributions and when they are considered in a probability sense,, they become *probability density functions* (PDF).

Another distinction must be made. For a probability mass function, it was possible to state the probability of occurrence at any *point* on the random variable scale. There are an infinite number of "points" on the random variable scale of a PDF ("There are an infinite number of points between any two points on a continuous scale"). If we have an infinite number of points, the probability under the curve of the PDF *at any point* is zero. So we cannot say the probability at point x is y but must say that the probability between points x_1 and x_2 is y.

The Gamma Distribution[1] The gamma distribution is a basic continuous statistical distribution which is onesided, i.e., the values of its random variable are bounded by 0 and $+\infty$. This distribution is used in queuing and reliability work because it represents the distribution of time required for exactly x independent events to happen given that the events happen at a constant rate. The gamma distribution is the parent distribution for the exponential, Erlangian, and chi-square distributions which are simply special cases of the gamma distribution.

The gamma probability density function is

$$f(x;\ n,\lambda) = \frac{\lambda^n}{(n - 1)!}\, x^{n-1} e^{-\lambda x} \tag{3.9}$$

where n = positive integer
x = a random variable
n = number of independent events to occur
λ = constant rate of occurrence
e = 2.718

For example, if a spare part is requisitioned in lots of size n and use of individual items occur independently and at a constant rate of λ per time unit, then the time between lot usage is a gamma-distributed variable.

Note that in (3.9) the function is restricted to n being a positive integer. Actually n can be a function, called the gamma

[1] This section may be omitted by the nonmathematically inclined.

function. In (3.9), $\Gamma(n) = (n - 1)!$. $\Gamma(n)$ is defined as

$$\int_0^\infty x^{n-1} e^{-r}\, dr \qquad (3.10)$$

Equation (3.10) is sometimes called Euler's integral. Note the r in the place of the usual x for clarity purposes when forms of the gamma distribution are involved. Model (3.10) is tedious to calculate, so it might be well here to pause and investigate the relationship between the gamma function and ordinary factorials.

Also, when n is restricted to integers, the gamma distribution is called the *Erlangian* distribution after the noted Danish statistician. The *chi-square* distribution, a special case of the gamma distribution when $\lambda = \frac{1}{2}$ and n is a multiple of $\frac{1}{2}$, will be covered later. This means that the distribution has only one parameter, v, which equals $2n$ and is, of course, always an integer. v is termed *degrees of freedom*. The exponential is a gamma with $n = 1$—but more about this distribution later.

To determine the most elementary properties of the gamma function and its connection with the factorial function

$$k! = k(k - 1) \cdots 3(2)(1)$$

apply integration by parts to (3.10), letting

$$u = e^{-r} \qquad dv = r^{n-1}\, dr$$
$$du = -e^{-r}\, dr \qquad v = r^n/n$$

Then

$$\Gamma(n) = \left.\frac{r^n e^{-r}}{n}\right|_0^\infty + \frac{1}{n}\int_0^\infty e^{-r} r^n\, dr \qquad (3.11)$$

If we restrict $r > 0$, the integrated part of (3.11) is undefined and therefore drops out at both limits. Comparing (3.11) with (3.10), it can be seen that the integral which remains is $\Gamma(n + 1)$ and the recursion relationship

$$\Gamma(n) = \frac{\Gamma(n + 1)}{n} \qquad (3.12)$$

results.

Equation (3.12) can also be written as

$$n\Gamma(n) = \Gamma(n + 1) \tag{3.13}$$

Computing $\Gamma(1)$ from (3.10),

$$\Gamma(1) = \int_0^\infty e^{-r}\,dr = -e^{-r}\Big|_0^\infty = 1$$

and using (3.13),

$$\begin{aligned}\Gamma(2) &= 1 \cdot \Gamma(1) = 1 \\ \Gamma(3) &= 2 \cdot \Gamma(2) = 2 \cdot 1 = 2! \\ \Gamma(4) &= 3 \cdot \Gamma(3) = 3 \cdot 2! = 3!\end{aligned}$$

and as a general case

$$\Gamma(k + 1) = k! \qquad k = 1, 2, 3, \ldots$$

Now back to the discussion of the gamma distribution per se.

Example: A cable car leaves as soon as all nine seats are filled. During any given time period, customers arrive independently at an average rate of 12 per hr. The owner wants to know the probabilities that the time between trips will be less than 1 hr.

In this problem we shall make use of a table of incomplete gamma-function ratios. Complete tables can be found in reference 1 at end of this chapter and an abbreviated table is included as Appendix 3.

In these tables we want to determine the probability that a value in the gamma distribution with parameters λ and n assumes a value less than r. To do this we must first compute the u ratio, which is defined as

$$u = \lambda r / \sqrt{n} \tag{3.14}$$

Additionally we need a parameter p defined as

$$p = n - 1 \tag{3.15}$$

Equations (3.14) and (3.15) enable us to get $I(u, p)$, which are the ratios that are needed to use the tables.

For the problem under discussion,

$$u = \lambda r/\sqrt{n} = 12(1)/\sqrt{9} = 12/3 = 4.0$$

and $p = n - 1 = 9 - 1 = 8$. Therefore, consulting the table in Appendix 3, we get 0.845. So approximately 15% of the time more than 1 hr will have elapsed between trips.

The mean of the gamma function is n/λ, and the variance is n/λ^2. Naturally, since the gamma function is dependent on n and λ, no single curve can describe the function. A typical family of curves is shown in Figure 3.5.

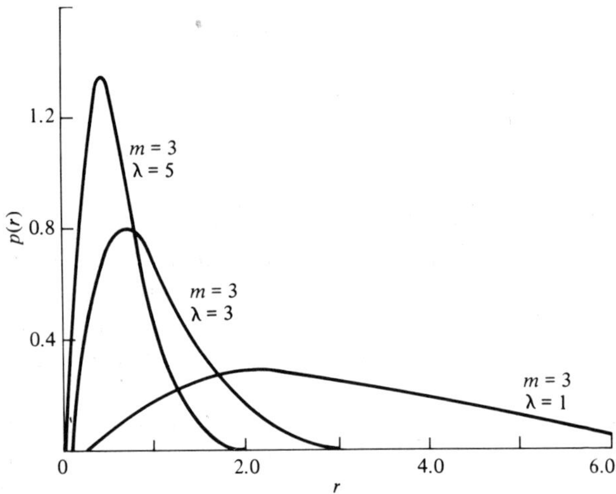

Fig. 3.5. Frequency distribution plots for the gamma distribution.

The Beta Distribution[2] The gamma distribution describes random variables bounded at one end, whereas the beta distribution can represent variables that function over an interval. The beta distribution can be used to express the distribution of yield in a manufacturing process in terms of percentage during some time interval. It also expresses the elapsed times to task completion in PERT.

---

[2]This section may be omitted by the nonmathematically inclined.

Generally the limits of the beta distribution are 0 and 1. The beta distribution is

$$f(r;\, v, n) = \frac{\Gamma(v + n)}{\Gamma(v)\,\Gamma(n)}\, r^{v-1}(1 - r)^{n-1}, \qquad 0 \le r \le 1 \quad (3.16)$$

where $v = n_s - s - l + 1$

n_s = sample size

s = degrees of freedom attendant with smallest observation

l = degrees of freedom attendant with largest observation

$n = s + l$

r = proportion (in percent) of population between highest and lowest values

Example: Thirty-five samples are chosen at random. The process is adjusted so that future samples will lie between the smallest and the largest value of the samples. What is the probability that at least 95% of a large number of future samples will fall between these values?

Tables for percentage points of the beta distribution are found in reference 1. From these tables, r (defined above) for $\alpha = 0.05$ and 34 and 2 degrees of freedom is 0.871. α, the "level of significance," will be explained shortly. Therefore the probability that at least 95% of the samples are within the sample range is 0.871. The beta distribution represents, as has been said, physical variables whose values are restricted to definite intervals. Further, the beta represents the estimated time to completion of phases of projects. In the PERT technique, the scheduler estimates the optimistic (o), pessimistic (p), and most likely (m) times required to complete each event. With this information a beta distribution over the interval (o, p), with a mode at m and a standard deviation equal to $\frac{1}{6}(p - o)$, can be assumed. Figure 3.6 illustrates a beta distribution applied to the PERT example.

When $v = n = 1$, the beta distribution becomes the *uniform* or *rectangular* distribution. When $v = 2$ and $n = 1$, the *right-triangular* distribution results. If $v = n = 2$, the *parabolic* distribution results.

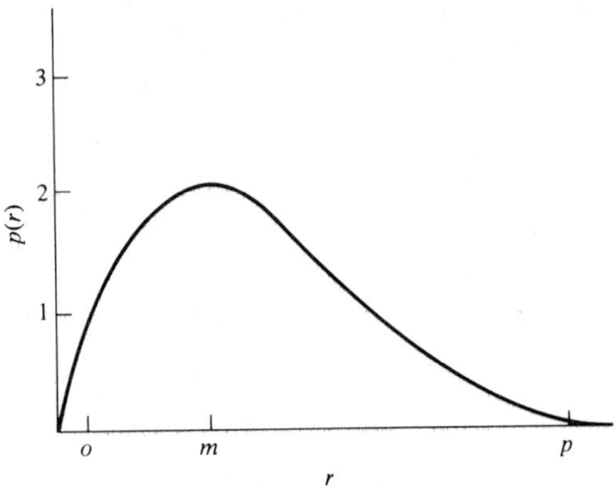

Fig. 3.6 Beta distribution as applied to PERT.

The Exponential Distribution[3] As already noted, when $n = 1$ in the gamma distribution, the resulting distribution is called the *exponential*. The exponential distribution is most important in reliability problems. Because of this, much maintenance theory depends on the exponential and the Weibull distributions.

The density function for the gamma distribution is

$$f(r) = \frac{\lambda^n}{(n - 1)!} r^{n-1} e^{-\lambda r}$$

and substituting $n = 1$ in the above expression we get

$$f(r) = \frac{\lambda^1}{(1 - 1)!} r^{1-1} e^{-\lambda r} = \lambda e^{-\lambda r} \tag{3.17}$$

since $0! = 1$ and $r^0 = 1$, and where λ is the average number of events per unit space or time, and r is the interval between events. Because the exponent on the e is negative, Eq. (3.17) is sometimes called the *negative exponential* distribution. In reliability work, the λ parameter is called the failure rate, and the time to failure is t, with the *mean* time to failure denoted by θ, which is $= 1/\lambda$. The parameter t corresponds to the more general r of Eq. (3.17) and others.

---

[3]This section may be omitted by the nonmathematically inclined.

If the distribution of failure times is an exponential one and a constant failure rate is assumed, we can state some reliability relationships.

To use the product probabilistic relationships, we have to get a relation expressing reliability in terms of t (see above). Using R for reliability (closely related to probability) we get

$$R(t) = 1 - f(t) = 1 - \int_0^t f(t)\,dt$$

from which we get

$$R(t) = 1 - \int_0^t \lambda e^{-\lambda r}\,dr = e^{-\lambda t} \tag{3.18}$$

which is the reliability expression for the exponential distribution.

Example: A number of shock absorbers were tested to failure and it was found that the failure rate was 0.04 per 100 hr. What is the probability that a shock absorber will survive at least 1000 hr of operation?

$$R(1000) = e^{-(0.04)\,10} = e^{-0.4} = 0.67$$

A table of positive and negative exponents of e is included as Appendix 2.

Since the exponential is a special case of the gamma distribution, it is not surprising that the mean equals $1/\lambda$ and the variance equals $1/\lambda^2$. Figure 3.7 shows a typical plot of the exponential distribution for θ's of 0.5, 1, and 2.

The Weibull Distribution[4] The Weibull distribution is applicable as a general case of the exponential, that is, when the failure rate is not constant. The Weibull distribution describes the failure occurrences of components when the failure rate either decreases or increases with respect to time. This, of course, adds another parameter, and the expression for the Weibull distribution is

[4]This section may be omitted by the nonmathematically inclined.

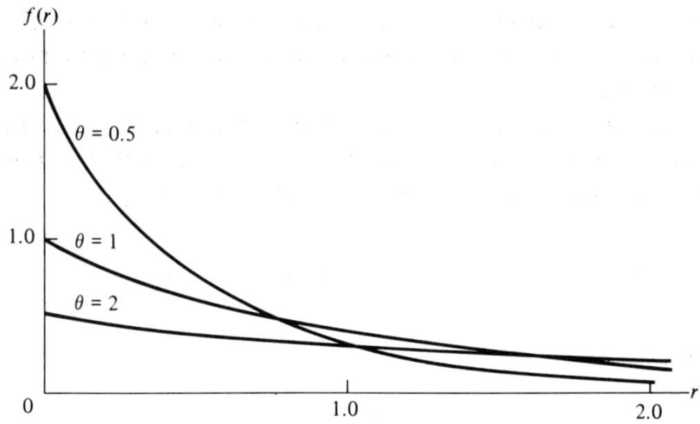

Fig. 3.7 Frequency distribution plots of the exponential distribution.

$$f(t) = \lambda b t^{b-1} e^{-\lambda t^b} \tag{3.19}$$

where t is a special case of the more general r, b is an experimentally determined parameter called the Weibull slope, and the other variables are as previously defined.

The mean of the Weibull is

$$\lambda^{-1/b}\Gamma\left(1 + \frac{1}{b}\right)$$

and the variance is

$$\lambda^{-2/b}\left\{\Gamma\left(1 + \frac{2}{b}\right) - \left[\Gamma\left(1 + \frac{1}{b}\right)\right]^2\right\}$$

Typical curves (for $\lambda = 1$ and various b's) are shown in Figure 3.8.

The Normal Distribution The normal distribution, the giant among statistical distributions, has been saved for the end of our discussion on the subject. Its application to engineering (as well as other) situations is so widespread as to defy limitations. The distribution in Table 2.2, graphed in Figure 3.1, is a normal one.

This curve was first recognized and used by Abraham de Moivre, a French Huguenot mathematician who lived in the

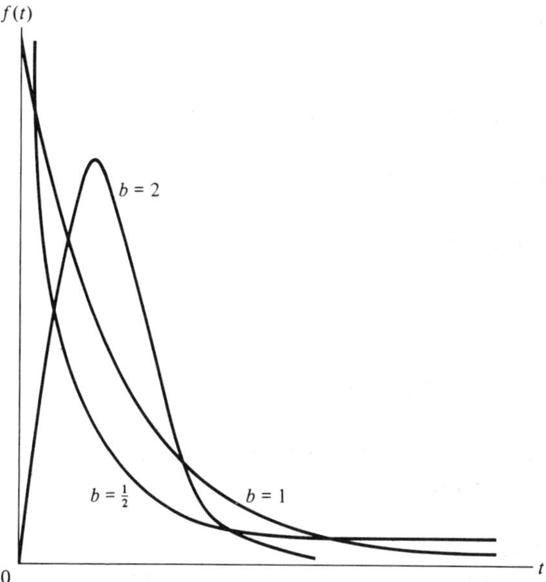

Fig. 3.8 Frequency distribution plots of the Weibull distribution.

seventeenth century. But it was not until the nineteenth century that Carl Friedrich Gauss developed and expressed the equation for the distribution. A very interesting accounting of the contributions of Gauss can be found in reference 2. The normal curve has many names of which a few are

The normal law
The normal curve of error
The Gaussian curve
The probability curve
The Laplacian curve
The normal distribution curve

Another reason for the importance of this statistical distribution lies in the *central limit theorem*. This theorem states that "if the mean μ and the variance σ^2 of *any* probability distribution are finite, then the distribution of the means of the samples approach the normal distribution with mean μ and variance σ^2/n as the sample size n increases."

The implication of the central limit theorem lies in the fact that we do not have to identify the probability distributions if it is possible for us to use the means of samples from those distributions.

Since the normal distribution is central to the engineering theme, many statistical functions relating to it have been developed and are widely used. It is these functions that are of most interest to us rather than the theoretical distribution as such.

The equation for the normal curve is

$$f(x) = \frac{1}{\sigma\sqrt{2\pi}} e^{\{[-(x-\mu)^2]/2\sigma^2\}} \tag{3.20}$$

where σ = standard deviation (defined in Chapter 2)
x = random variable interval
μ = mean (defined in Chapter 2)
π = 3.142
e = 2.718

Of course, the mean of the distribution is μ and the variance is σ^2. Figure 3.9 illustrates the curve for the normal distribution.

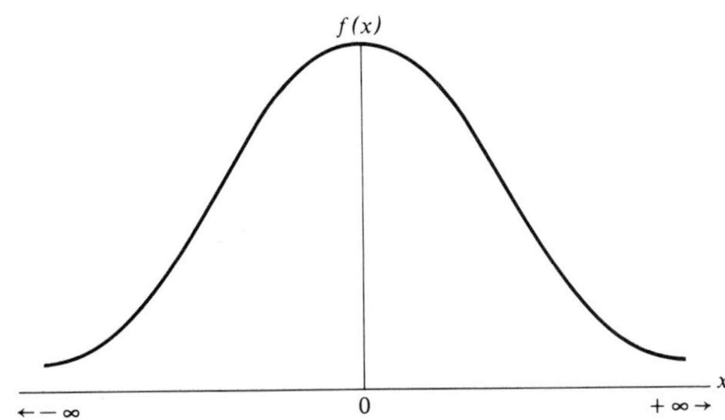

Fig. 3.9 Frequency distribution plot of the normal distribution.

The total area under the curve, from $-\infty$ to $+\infty$, equals 1.0. Since this is a continuous function, the relative frequency of occurrence of any variable between, say, x_1 and x_2 is measured by the fraction of area under the curve multiplied by the total population, N. The area between any two values on the random variable scale expressed as a probability is

$$p(x_1 \leq x \leq x_2) = \frac{1}{\sigma\sqrt{2\pi}} \int_{x_1}^{x_2} e^{-[(x-\mu)^2/2\sigma^2]}\, dx \qquad (3.21)$$

Similarly, the probability of observing a value of a variable which is normally distributed equal to or less than a value x_1 is the area under the curve from $-\infty$ to x_1, which is

$$p(x \leq x_1) = \frac{1}{\sigma\sqrt{2\pi}} \int_{-\infty}^{x_1} e^{-[(x-\mu)^2/2\sigma^2]}\, dx \qquad (3.22)$$

and, of course,

$$p(x \geq x_2) = \frac{1}{\sigma\sqrt{2\pi}} \int_{x_2}^{+\infty} e^{-[(x-\mu)^2/2\sigma^2]}\, dx \qquad (3.23)$$

Now consider the random variable scale of the theoretical normal distribution [model (3.20) and Figure 3.9]. We must express that random variable in some unit. When this variable is expressed in standard deviation units with the mean at zero, we get a standardized normal distribution which is denoted by $N(0, 1)$. This notation says, in effect, "a normally and independently distributed variable with mean 0 and variance 1." The standard measure, or standard deviate, as it is called, is denoted by z. We are actually expressing each value of x in terms of the number of standard deviations that it varies from the mean: $x = \mu + z\sigma$, or

$$z = \frac{|x - \mu|}{\sigma} \qquad (3.24)$$

You will recognize this as a transform, if we substitute $\mu + z\sigma$ for x in Eq. (3.22), noting that μ and σ are constants. If

$$x = \mu + z\sigma$$

then

$$dx = z\, d\sigma + \sigma\, dz$$

and the $z\, d\sigma$ term goes to zero since the differential of a constant is zero. Therefore $dx = \sigma\, dz$ and, as said above, $x = \mu + z\sigma$. Therefore, substituting in (3.20), we get

$$f(x)\,dx = \frac{1}{\sigma\sqrt{2\pi}}\, e^{[-(\mu+z\sigma-\mu)^2]/2\sigma^2}\, \sigma\, dz$$

$$= \frac{1}{\sqrt{2\pi}}\, e^{-z^2/2}\, dz \qquad (3.24a)$$

A graph of this standardized normal curve is shown as Figure 3.10.

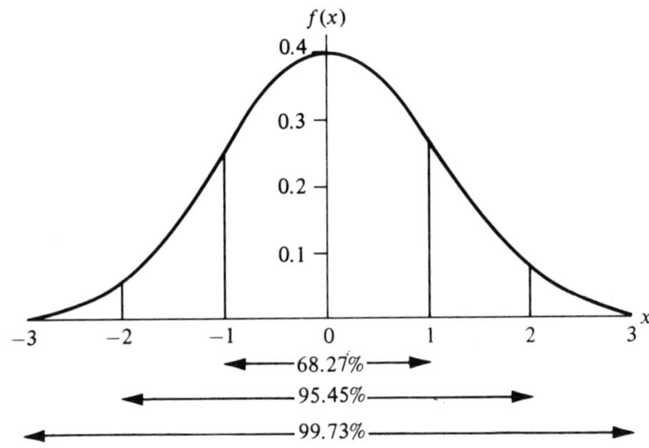

Fig. 3.10 Normal distribution frequency plot showing the 1,2 and 3σ limits.

In this curve, the areas included between $z = -1$ and $+1$, $z = -2$ and $+2$, and $z = -3$ and $+3$ are shown together with the attendant respective percentages of the area of the total curve.

Let us dwell on these properties for a moment. We have said, in effect, that whenever we have a normal distribution and the sample sizes are large enough, we can expect certain percentages within certain intervals on the random variable scale. This is the essence of control, in general, and statistical quality control in particular. Note the relationship between variance and the z value. If we are attempting to control an industrial process, it would be unrealistic to expect exactly the same results from every run, even though no physical variables change. How much can we let the process vary and still call it "in control?" The answer to this question lies in an appreciation of the natural, inherent variability of the process and how much of our yield we will tolerate being "out of specs." The z value quantifies these limits for us.

We see from Figure 3.10 that 68.27% of the area under the curve (equivalent to 68.27% of the samples) lies between $\pm 1z$, 95.45% within $\pm 2z$, etc. Of course, we do not have to deal with integer values of z. The table in Appendix 4 gives the area under the normal curve for various values of z. These values represent the probabilities for normally distributed variables of the occurrence between $-\infty$ and some z. The table in Appendix 4 lists all values of z with the minus sign since we are concerning ourselves only with the left half of the normal curve. This is possible because of the symmetry of the function.

To become familiar with the z table, consider the following examples.

Example: Find the area under the normal curve between $z = -\infty$ and $z = -1.23$.

In these types of problems, confusion can be avoided if the distribution and z points are plotted as shown in Figure 3.11.

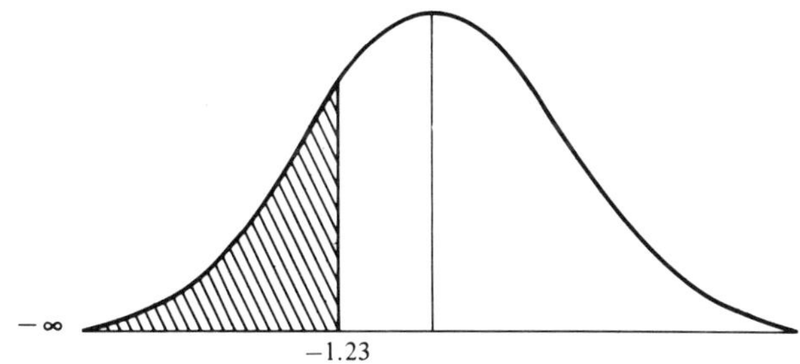

Fig. 3.11 z distribution for example opposite.

From Appendix 4, the z value for -1.23 is 0.1093. Therefore 10.93% of the area lies to the left of $z = -1.23$.

Example: Find the area under the normal curve between $z = -\infty$ and $z = 1.51$.

From symmetry we know that the area from $-\infty$ to 0 is 0.5000. The area from $z = 0$ to $z = 1.51$ is the same as the area from $z = -1.51$ to $z = 0$. From the table the area of $z = -1.51$ is 0.0655; therefore $0.5000 - 0.0655 = 0.4345$. But we want the area from $-\infty$ to 1.51. So adding 0.5000 + 0.4345, we get 93.45% of the area. Of course, the correct answer could have been obtained by subtraction: 1.000 − 0.0655 = 0.9345. See Figure 3.12.

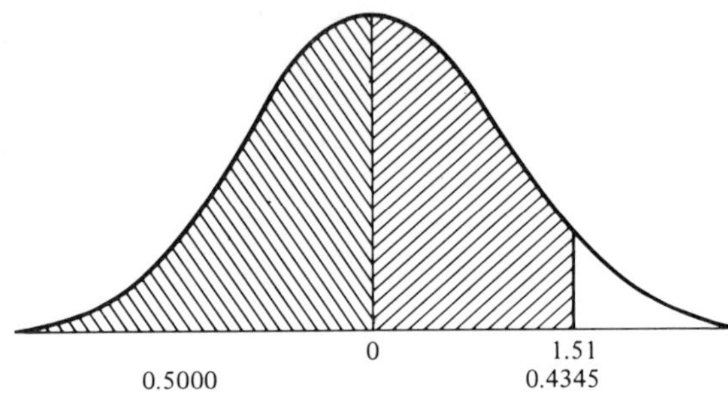

Fig. 3.12 z distribution for example opposite.

Statistical Hypotheses By discussing the z distribution, we have entered the area known as *statistical decision theory*, which is, as the name implies, decision based on statistical treatment of the input data. To reach these decisions we make assumptions or guesses about the statistical data involved. We may make one assumption hoping to prove it false or make another assumption hoping to prove it true. Both assumptions involve the same problem, and we can gain our end via either route. These assumptions are generally statements about the probability distributions of the populations.

For example, if we want to prove that a pair of dice is not loaded, we formulate the hypothesis that the probability of "shooting" a 7 is $\frac{6}{36}$; i.e., $p(7) = \frac{1}{6}$. Frequently we want to decide whether one process is the same as another, or whether one process is better than another, or whether one process is different from another.

When our hypothesis is one of "no difference," then we call this the *null* hypothesis and denote it by H_o. A hypothesis that differs from the given hypothesis is called the alternative hypothesis and is denoted by H_A.

Naturally, in stating any hypothesis, we are not absolutely sure that an event will or will not take place. We are absolutely sure of very few events. Once having applied a statistical test to the hypothesis in question, we must indicate the degree to which we are confident of the result of the test. This degree is a probability denoted by α and called the *level of significance*. In most engineering applications a level of significance of 0.10, 0.05, or 0.01 is used. For example, if 0.05 were chosen (sometimes called a 5% level of significance), there would be a 5% chance of making the wrong decision, i.e., of rejecting the hypothesis when it should be accepted. We can also say that we are 95% confident of being right. The 95% is the *confident limit*. The distinction between level of significance and confidence limit is subtle, but important, as we shall see.

In the z test, which has been discussed, and in the t test, which we shall get to shortly, we are interested in areas either outside or inside some limits which partitioned the normal distribution curve. In Figure 3.13 we might be interested in

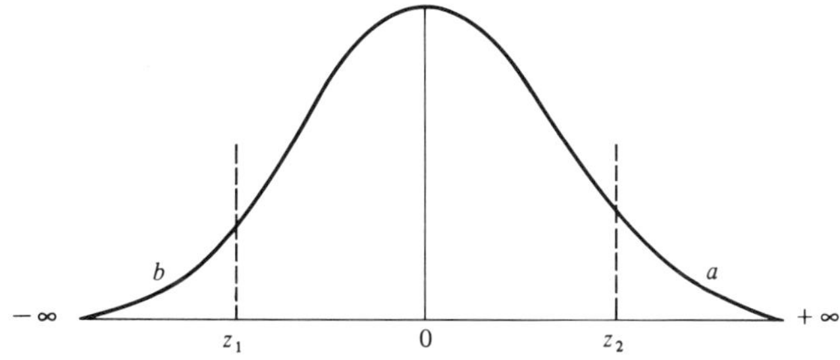

Fig. 3.13 z distribution illustrating two-tail test.

the area between z_1 and z_2. If this were so and we obtained, say, 90% of the area between z_1 and z_2, then 10% would be outside, i.e., 5% in *a* and 5% in *b* due to symmetry. This is known as a *two-tailed test*. Conversely, if we were interested in a test which would indicate whether one process was better (or worse) than another, and if we were asked to decide with a yes or no answer, then "considerably better" versus "somewhat better" would not be very helpful except that we could make our statement with more confidence. This would be a *one-tailed test*.

Similarly, if we were concerned with a one-tailed test and 10% were outside our area of interest, then all 10% would be concentrated in areas *a or b* in Figure 3.13. If we were interested in a two-tailed test, the entire 10% would be concentrated in areas *a and b* (5% in each side if symmetrical).

Let us return to our consideration of hypotheses. Assume that we have a sample from a productive process and we want to compare it with successful past runs of the same process. Let μ_p be the mean of past successful runs and μ_1 be the mean of the run in question. Also, let us formulate our hypothesis as H_o: $\mu_1 = \mu_p$, which you recognize as the null hypothesis. The alternative hypothesis would be H_A: $\mu_1 \neq \mu_p$. When we stated the null hypothesis we assumed that the sample was from the large population. The second assumption H_A was that the population was different from the sample population or, stating it another way, that the null hypothesis was not true. The null and alternative hypotheses are mutually exclusive. If, as a design criteria, we choose an α, the α is the probability of making a mistake by saying that the null hypothesis is false when it is really true. This type of error is called a type I error. $1 - \alpha$ is called the *confidence level* and is the probability of making the right decision when the null hypothesis is true.

To complete the process we must find the probability β of accepting the null hypothesis when the alternative hypothesis is true. An error of this type is called a type II error. If α and β are not small enough, we can say only that there is not enough evidence to draw a conclusion that is statistically significant. $1 - \beta$ is called the *power of the test* and is defined as the probability of rejecting the hypothesis when it is false.

Illustrating all this graphically, consider a test in which we assume that the mean of a normal distribution has a specified value. The hypothesis would be H_o: $\mu_1 = \mu_p$. In Figure 3.14 two distributions are shown, that for the entire population,

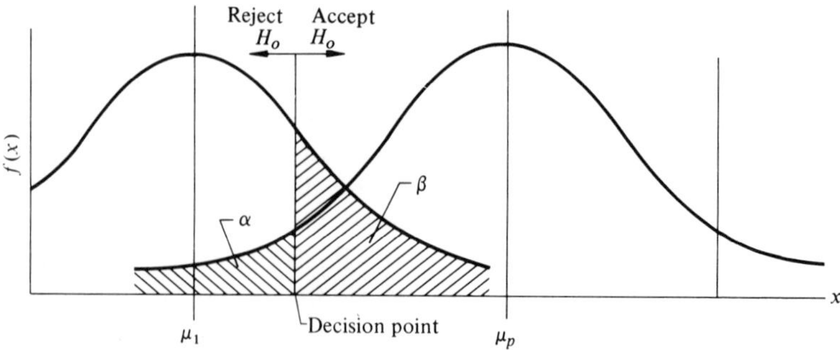

Fig. 3.14 One-sided test.

with mean μ_p, and that of the sample, with mean μ_1. The hypothesis is true if there is no significant difference between the means; hence $\mu_1 = \mu_p$. If the sample population actually does come from the larger population, it must be detected with probability $1 - \beta$.

The type I error, α, is the error of saying that the sample belongs to a distribution centered at some value less than μ_p when it is actually centered at μ_p. The α error is the area under the population distribution to the left of the decision point. The *decision point* is determined by the sample size and the chosen α. The numerical examples will more fully explain the decision point. The type II error, β, is the probability of stating that the sample population came from the larger population when actually it came from the distribution to the left.

Note that, if the decision point is moved to the left, the area β will be larger and the chances of making a type II error will be increased. At the same time the alternative will have a reduced chance of being accepted as true.

The above arguments hold true for the two-sided test, shown in Figure 3.15. The type I error, α, is the error of saying

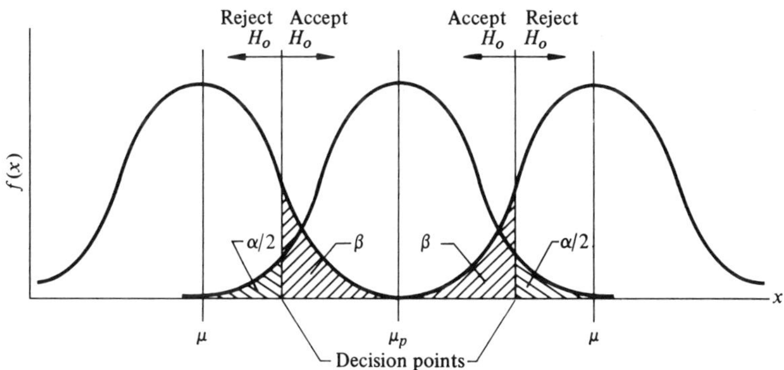

Fig. 3.15 Two-sided test.

that the sample distribution does not coincide with the population distribution when it actually does. This error is encountered by finding a sample mean which varies from μ_p either to the right or left. Therefore, the probability of making this error is apportioned equally and hence $\alpha/2$ in each of the two tails. The type II error, β, occurs if H_o is accepted when the sample distribution coincides with either of the distributions with mean μ. Therefore the areas β represent the probability of making such an error.

The *t* Test The t test entails the estimating of a true value from a sample and establishing a confidence range within which we can say that the true value lies. We described the z function as a measure of the deviation of a variable from some mean in terms of the standard deviation:

$$z = \frac{|x - \mu|}{\sigma}$$

Noting the greek letters, in the z distribution we have the universal population mean and standard deviation. The z distribution is limiting from a number of standpoints. First, it is a distribution of a theoretical normal situation, and this is seldom prevalent unless we have a very large sample size. Second, it is assumed that we know the population mean and standard deviation. Again, we seldom have these parameters available to us. The t distribution, and it is a distribution, can be defined in a manner similar to the z distribution. It is the difference between the mean of some sample and the true mean of some population from which the sample may or may not have been drawn, divided by an *estimate* of the standard deviation of the mean. t is defined as

$$t = \frac{|\bar{x} - \mu|}{s(\bar{x})} \tag{3.25}$$

It is a nonnormal distribution with a mean at zero and a variance of $\nu/(\nu - 2)$ for $\nu > 2$, where the ν's are degrees of freedom. The density function for the distribution is

$$p(x) = \frac{1}{\sqrt{\nu\pi}} \frac{\Gamma[(\nu + 1)/2]}{\Gamma(\nu/2)} \left(1 + \frac{t^2}{\nu}\right)^{-\nu + 1/2} \tag{3.26}$$

A plot of the distribution is shown in Figure 3.16. The t distribution is more correctly called Student's t distribution. It was developed in 1907 by W. S. Gosset, who used the pseudonym *Student* because the brewery in England where he worked discouraged employees from publishing where the pos-

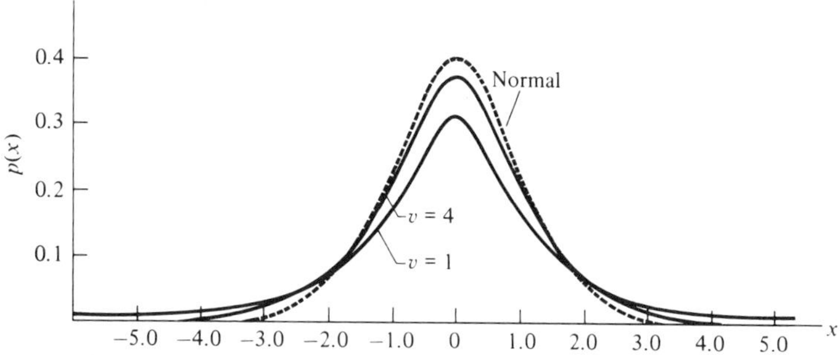

Fig. 3.16. Plot of Student's t distribution.

sibility of revealing trade secrets existed. R. A. Fisher, a statistician of considerable note, further refined the distribution.

When n is the sample size (from 2 to ∞), then t has a different value for each value of n. As n approaches ∞, the t distribution approaches the normal distribution. Again see Figure 3.16.

Consulting model (3.25), it is not readily apparent how n affects the distribution. This occurs in the $s(\bar{x})$ term. $s(\bar{x})$ is the estimated standard deviation of the means of samples of n size. These samples are taken from a population which is estimated to have a standard deviation of $s(x)$. In manipulating data for use in a t test, the following relationship is most important and the student should not pass over it lightly:

$$s(\bar{x}) = s(x)/\sqrt{n} \qquad (3.27)$$

The values of t for some degrees of freedom will be found in Appendix 5, together with the probability levels for observing larger absolute values than those shown. This table is the first of three (t, chi-square, and F) which have limiting values. In other words, the values in the tables are the upper limit of the t distribution, for a given n and α, where there is no difference between the sample and the population distribution. When the calculated t does not exceed the t in the table, we cannot say that there is a difference *between the means*.

For an example of using the table, with 7 deg of freedom, there is only a 0.05 probability of observing a t larger than

2.36 or smaller than −2.36. Bear in mind that the 5% includes the area to the right of 2.36 and to the left of −2.36.

The *t* test can be formulated in a number of ways depending on the hypothesis. One of the most common formulations is used when a decision is to be made as to whether the sample mean m is truly a sample from a larger population with mean μ. The hypothesis is H_o: $\mu = m$.

At this point it is necessary to introduce the *sums of squares* concept, usually denoted by SS. Literally this notation means "sums of squares of deviations from the mean." This is not entirely new to us since we encountered the idea when calculating the standard deviation. Therefore the quantity $\Sigma(x - \bar{x})^2$ is referred to as SSx^2 and from model (2.8) we have

$$SSx = \Sigma x^2 - \bar{x}\,\Sigma x = \Sigma x^2 - \frac{(\Sigma x)^2}{n} = (n - 1)\,s^2(x) \quad (3.28)$$

Example: A connecting rod for a small engine is supposed to be 2.750 in. long. The data shown in Table 3.4 are the measurements of 13 rods. Test at the 0.05 level to ascertain whether these data might come from a population with a mean of 2.750 in.

TABLE 3.4

Connecting-Rod Sample Measurements
2.74
2.63
2.64
2.68
2.79
2.81
2.80
2.72
2.78
2.74
2.79
2.69
2.70

$$
\begin{aligned}
n &= 13 \\
\Sigma x &= 35.51 \\
\Sigma x^2 &= 97.04 \\
\bar{x} &= \Sigma x/n = 35.51/13 = 2.732 \\
\mathrm{SS}x &= \Sigma x^2 - \bar{x}\,\Sigma x = 97.04 - 2.732(35.51) = 0.03
\end{aligned}
$$

and from (3.28) we get

$$s^2(x) = \mathrm{SS}x/n - 1 = 0.03/12 = 0.0025$$

Therefore from (3.27)

$$s^2(\bar{x}) = s^2(x)/n = 0.0025/13 = 0.0002$$

and

$$s(\bar{x}) = 0.014$$

The hypothesis is H_o: $\mu = m$, or, in terms of the problem, H_o: $\bar{x} = m$. So from (3.25)

$$t = \frac{|\bar{x} - m|}{s(\bar{x})} = \frac{2.732 - 2.750}{0.014} = \frac{0.018}{0.014} = 1.285$$

Because there are 13 sample points, there are 12 deg of freedom. The tabular value of t with 12 deg of freedom and an α of 0.05 is

$$t_{0.05,12} = 2.18$$

Since the calculated value does not exceed the tabular value for the chosen significance level, then, based on this data, we accept the hypothesis that our data are, in fact, from the larger population. In effect, we are saying that with the variability that we have we cannot determine any difference between 2.750 and 2.732. The student should realize that if either the difference in the numerator of the t expression were greater or the scatter [as reflected by $s(\bar{x})$] were smaller, then the calculated t would be higher, and at some point we would exceed the tabular t and we would reject the hypothesis (therefore find a difference).

Also note from the table that as α gets greater for a given number of degrees of freedom, the t values decrease. Since α is the probability of making an error, then as that probability goes up, our limiting value goes down; and, conversely, as our significance level of error gets smaller, our limit is broader. What we are saying is that if we want to make a statement with increasing accuracy, then we have to have a broader range so that we can be more sure that the range includes data results.

The One-Sided *t* Test The t test previously considered made use of the absolute value in the numerator. Appendix 5 gives only positive values for t but, due to symmetry (see Figure 3.16), positive and negative values equally apply. A value of t of 2.20 for 11 deg of freedom and an α of 0.05 indicates that there is a 0.025 probability of obtaining a t value less than -2.20 and, similarly, a 0.025 probability of obtaining a t value greater than 2.20. This concept is not new; it is merely an application of the one-sidedness principle to the t test. This type of one-sided test involves only the probability of occurrence in one-half of the t distribution, which is half the probability found in Appendix 5.

Naturally, the hypotheses which we state for a one-sided test are different from the two-sided ones. We are now dealing with inequalities; that is, rather than stating that some population mean μ is equal to some sample mean m, we say that the true mean is *not less* or *more* than some specific value. This can be expressed

$$H_o\colon \mu \leq m$$

and the t expression would be

$$t = \frac{m - \bar{x}}{s(\bar{x})} \tag{3.29}$$

Remember, if an α of 0.05 is the criteria, the hypothesis is rejected if the calculated value of t exceeds the 0.10 level because the tabular 0.10 level corresponds to the 0.05 probability on either side of the distribution. As before, a schematic sketch of the distribution, identifying each area, will help.

Example: The heat treatment given a certain type of spring is claimed to increase the yield strength of the metal by more than 5%. A number of springs were subjected to the treatment with the results shown in Table 3.5. The hypothesis is H_o: $\mu \leq m$. We establish our α level at 0.05. Therefore we will compare our t with a tabular t at the 0.10 level. Determine whether or not there is a difference between the sample and the larger population.

TABLE 3.5 Yield Strength of Spring Steel

Test Number	Percent Increase in Yield Strength
1	7.1
2	7.4
3	4.6
4	8.1
5	5.3
6	3.7
7	4.3
8	5.9

$$
\begin{aligned}
n &= 8 \\
\Sigma x &= 46.4 \\
\bar{x} &= 5.80 \\
\Sigma x^2 &= 287.02 \\
SSx &= \Sigma x^2 - \bar{x}\,\Sigma x = 287.02 - 5.80(46.4) = 17.90 \\
s^2(x) &= SSx/(n-1) = 17.90/7 = 2.557 \\
s^2(\bar{x}) &= s^2(x)/n = 2.557/8 = 0.320 \\
s(\bar{x}) &= 0.566 \\
t &= \frac{5.80 - 5.00}{0.566} = \frac{0.800}{0.566} = 1.41 \\
t_{0.10,7} &= 1.90
\end{aligned}
$$

Since the tabular t value is greater than the calculated value, we accept the hypothesis that the actual mean increase in strength is not more than 5%.

Comparison of Paired Tests One of the most common tests is that of determining whether one process is different from another based on comparisons between pairs of values. The closer to zero that the total of the differences lies, the more confident that there is no difference between the processes. Saying this another way: Is the total of the differences significantly different from zero? Hence, the null hypothesis is

$$H_o: \mu = 0$$

Since we are testing the difference between the two values that make up the pair, we are not interested in the differences in the totals of all tests on process 1 and all tests on process 2. In this way the tests can be arranged in ascending, descending, or even random order, and, so long as each pair is made together, the measurements are valid for our purposes. Note that this is a one-tailed test. In the parlance of experimental design, we are looking for differences *within* pairs, not differences *between* pairs. This distinction will be considered in more detail later. The number of degrees of freedom remains $n - 1$, where n here is the number of pairs, i.e., the number of measurements of difference.

Example: Table 3.6 shows the coded comparison between successive samples run on different mass spectrometers analyzing the same gas. Using the t test, determine whether there is a significant difference between the spectrometers, using an α of 0.05. Using the data given, which spectrometer is more accurate?

TABLE 3.6 Mass Spectrometer Results of Gas Analysis

Sample	Spectrometer 1, x_1	Spectrometer 2, x_2	α $(x_1 - x_2)$
1	4	6	−2
2	5	5	0
3	7	8	1
4	9	7	2
5	8	6	2
6	18	14	4
7	17	13	4
8	20	15	5
9	21	18	3

$$n = 9$$
$$\Sigma\, d\,(\text{algebraic}) = 19$$
$$\bar{d} = 2.111$$
$$\Sigma\, d^2 = 79$$
$$\text{SS}d = \Sigma\, d^2 - \bar{d}\,\Sigma\, d = 79 - 2.111(19) = 38.891$$

$$s^2(d) = \mathrm{SS}d/(n - 1) = 38.891/8 = 4.861$$
$$s^2(\bar{d}) = 4.861/9 = 0.540$$
$$s(\bar{d}) = 0.736$$
$$t = \frac{|2.111 - 0|}{0.736} = 2.86$$
$$t_{0.05,8} = 2.31$$

Since the calculated t is more than the tabular t, we reject the hypothesis of no difference and conclude that the two spectrometers are different.

Which of the spectrometers is more accurate? There is no way for us to determine this with the data at hand.

Note that if we had used an α of 0.01, the tabular t would have been 3.36 and our conclusion would have been different. The reason for this lies in the fact that we would be asked to make the statement with considerably less chance of error and the data would not have supported such a statement.

Test for Difference Between Two Means This test is different from the first t test that we discussed. There we were comparing a sample population against (for all practical purposes) an infinitely large population. The test to be discussed here concerns itself with the comparison of two sample, finite populations, not necessarily of equal sample size. The test actually has to do with determining whether the means of two different samples did, in fact, come from the same population with the same means. Hence, the null hypothesis:

$$H_o: \mu_1 = \mu_2$$

To determine the calculated t, we must discuss a parameter called the *pooled estimate of the standard deviation*, which is denoted by $\bar{s}(x)$ and is defined by the following model:

$$\bar{s}(x) = \sqrt{\frac{\mathrm{SS}x_1 + \mathrm{SS}x_2 + \cdots + \mathrm{SS}x_m}{n_1 + n_2 + \cdots + n_m - m}} \qquad (3.30)$$

For the special case where $m = 2$, we get

$$\bar{s}(x) = \sqrt{\frac{\mathrm{SS}x_1 + \mathrm{SS}x_2}{n_1 + n_2 - 2}} \qquad (3.31)$$

We can now define the t test for the difference between two means as

$$t = \frac{|\bar{x}_1 - \bar{x}_2|}{\bar{s}(x)\sqrt{(1/n_1) + (1/n_2)}} \qquad (3.32)$$

Example: The data in Table 3.7 depict the coded results of tests performed on two tensile testing machines.

TABLE 3.7 Tension Testing Machine Results

	Tensile Strength	
Test Number	Machine 1, x_1	Machine 2, x_2
1	0.5	1.9
2	0.9	1.7
3	1.6	1.9
4	1.5	2.4
5	0.7	1.6
6	0.3	1.2

Determine, at an α level of 0.05, whether there is a difference between the test results.

$$H_o: \bar{x}_1 = \bar{x}_2$$
$$n_{x_1} = n_{x_2} = 6$$
$$\Sigma x_1 = 5.5; \qquad \Sigma x_2 = 10.7$$
$$\bar{x}_1 = 0.91; \qquad \bar{x}_2 = 1.78$$
$$\Sigma x_1^2 = 6.45; \qquad \Sigma x_2^2 = 19.87$$
$$\mathrm{SS}x_1 = \Sigma x_1^2 - \bar{x}_1 \Sigma x_1 = 6.45 - 0.91(5.5) = 1.44$$
$$\mathrm{SS}x_2 = \Sigma x_2^2 - \bar{x}_2 \Sigma x_2 = 19.87 - 1.78(10.7) = 0.82$$
$$s^2(x_1) = \mathrm{SS}x_1/(n_{x_1} - 1) = 1.44/5 = 0.29$$
$$s^2(x_2) = \mathrm{SS}x_2/(n_{x_2} - 1) = 0.82/5 = 0.16$$

From (3.31),

$$\bar{s}^2(x) = \frac{1.44 + 0.82}{10} = 0.226$$

so

$$\bar{s}(x) = 0.475,$$

and from (3.32),

$$t = \frac{|0.91 - 1.78|}{0.475\sqrt{(1/6) + (1/6)}} = 3.17$$

It is important to note that the degrees of freedom associated with this t test are $n_1 + n_2 - 2$, or 10 in this case. Therefore from the tables

$$t_{0.05,10} = 2.23$$

From this, we reject the hypothesis of equality and conclude that there is a difference between means.

Confidence Ranges It has been mentioned previously that parameters obtained from samples are simply estimates of true parameters. Further, we have begun to realize that our confidence in making statements is dependent on the sample size n and the significance level α. Earlier we learned that the smaller the scatter, the "better" the data. These ideas have been consolidated into the concept of confidence ranges, which have a wide use in engineering application. We realize that it is unrealistic to take a small sample, compute a mean $\bar{x}$, and then claim that $\bar{x}$ is the mean of the population. Our immediate question is: How far from $\bar{x}$ can we vary and still assume that the mean is indicative of our population? You immediately recognize this as within the realm of the t test. Realizing that,

$$p[\bar{x} - ts(\bar{x}) < \mu < \bar{x} + ts(\bar{x})] = 1 - \alpha \tag{3.33}$$

which says that the true mean is between the sample mean plus t times the standard deviation and minus t times the standard deviation. Equation (3.33) can also be read as the probability, before the sample is drawn, that the random interval is $\bar{x} \pm ts(\bar{x})$, or, stating the complete expression for confidence interval,

$$\mu = \bar{x} \pm ts(\bar{x}) \tag{3.34}$$

Example: We are interested in the mean weight of a ton of material. We assume the distribution to be a normal one. A random sample of the following values in pounds was taken:

2006
2008
2001
2003
2007
2005

The average of these values is 2005 lb, which is a reasonable estimate of the weight. But, we do not have any quantitative measure of the degree of confidence in this estimate. Use a 95% confidence limit ($\alpha = 0.05$).

x_i	$\bar{x}$	$x_i - \bar{x}$ (coded)	$(x_i - \bar{x})^2$
2006	2005	1	1
2008	2005	3	9
2001	2005	−4	16
2003	2005	−2	4
2007	2005	2	4
2005	2005	0	0
			34

$$\therefore s^2(x) = \frac{\Sigma (x_i - \bar{x})^2}{n - 1} = \frac{34}{5} = 6.80$$

$$s^2(\bar{x}) = 6.80/6 = 1.133$$

$$s(\bar{x}) = 1.065$$

$$t_{0.05,5} \text{ from Appendix 5} = 2.571$$

Therefore, from (3.34)

$$\begin{aligned}\mu &= \bar{x} - ts(\bar{x})\\ &= 2005 - 2.571(1.065)\\ &= 2002.26\end{aligned}$$

and

$$\begin{aligned}\mu &= \bar{x} + ts(\bar{x})\\ &= 2005 + 2.571(1.065)\\ &= 2007.74 \text{ lb}\end{aligned}$$

From this, it can be said that we are 95% confident that the true population mean weight is being 2002.26 and 2007.74.

If we had been interested only in the lower limit, we would have had a one-tailed test. All figures would be as above except the α for the t value. We want the entire 0.05 to be on one side rather than 2.5% on each tail. To get this we look up $t_{0.10,5}$ in Appendix 5, getting 2.02; therefore

$$\begin{aligned}\mu &= \bar{x} - ts(\bar{x})\\ &= 2005 - 2.02(1.065)\\ &= 2002.85 \text{ lb}\end{aligned}$$

Rejection of Unwanted Observation A special case of the t distribution is Chauvenet's criterion. This method gives us a decision-making tool to consider the rejection of outlier observations. The criterion states, "An observation in a sample of size n is rejected if it has a deviation from the mean greater than that corresponding to a $\frac{1}{2}n$ probability." We assume a normal distribution and an estimate of variance based on the sample to calculate the probability; that is, when $n = 10$, then $\frac{1}{2}n = 0.05$. If $n = 20$, then $\frac{1}{2}n = 0.025$. 0.05 and 0.025 are probabilities of deviates of some xs. Therefore a measurement which deviates from the mean by at least xs would be rejected. Table 3.8 gives the p (ratio of the deviation to the standard deviation) values (which correspond to the t tables) for various sample sizes. The example which follows will illustrate the use of the method.

TABLE 3.8 ***n* Versus *p***

n	p	n	p	n	p	n	p
2	1.15	7	1.80	15	2.13	50	2.58
3	1.38	8	1.86	20	2.24	100	2.81
4	1.54	9	1.91	25	2.33	250	3.09
5	1.65	10	1.96	30	2.40	500	3.29
6	1.73	12	2.04	35	2.45	1000	3.48

Rules for the use of the criterion are as follows:

1. Compute the mean and the standard deviation of *all* observations.
2. Determine the ratio of the suspiciously large deviation divided by the standard deviation.
3. From Table 3.8 determine the limiting value of that ratio (in item 2) for the corresponding number of determinations, n.
4. If the observed ratio is greater than the value found in the table, the observation may be rejected.

Example: Consider the following five readings obtained during a compressor run.

123.4
118.9
132.9
124.3
119.2

Should the 132.9 reading be discarded? The mean of these five values is 123.7 and the standard deviation is 5.07. The ratio of the suspected deviation to the standard deviation is

$$\frac{132.9 - 123.7}{5.07} = 1.81$$

For five determinations the tabulated value is 1.65. Because this is smaller than the ratio of the deviation to the standard deviation, the observation 132.9 should be rejected. The best value for the test would then be the mean of the remaining four determinations.

References

1. H. L. Harter, *New Tables of the Incomplete Gamma-Function Ratio and of Percentage Points of the Chi-Square and Beta Distributions*, Aerospace Research Laboratories, U.S. Air Force, U.S. Government Printing Office, Washington, D.C., 1964.
2. David Bergamini, *Mathematics* (*Life Science Library*), Time Inc., New York, 1963.

4
Chi-Square Tests

When considering distributions, it was noted that the chi-square distribution is the special case of the gamma distribution where $\lambda = \frac{1}{2}$ and n is a multiple of $\frac{1}{2}$. It was further noted that the degrees of freedom v were $= 2n$. Specifically, the chi-square distribution is the distribution of a random variable which consists of the sums of squares of other random variables. Assume there exists a situation consisting of γ trials and let $x_1, x_2, \ldots, x_\gamma$ denote the measurable quantity associated with each of the trials. The x's then are random variables which are normally and independently distributed with a mean of 0 and a variance of 1. Now let the sum of squares of these random variables $= \chi^2$ such that $\chi^2 = x_1^2 + x_2^2 + \cdots + x_\gamma^2$.

The density function of the chi-square random variable is

$$f(\chi^2) = \frac{1}{2^{\gamma/2}\Gamma(\gamma/2)} (\chi^2)^{(\gamma-2)/2} e^{-\chi^2/2} \tag{4.1}$$

for all positive values of χ^2 and 0 otherwise, where γ is the degrees of freedom.

Figure 4.1 shows the shape of the χ^2 distribution for some values of γ. Values of the χ^2 distribution for various α levels and degrees of freedom are included as Appendix 5.

The student will notice that in this chapter we are dealing with variance of counted or enumerated data, whereas in the t test we were dealing with differences between means which were measured rather than counted.

The relationship between χ^2 and s^2 is

$$s^2 = \frac{\sigma^2\chi^2(n-1)}{n-1}, \tag{4.2}$$

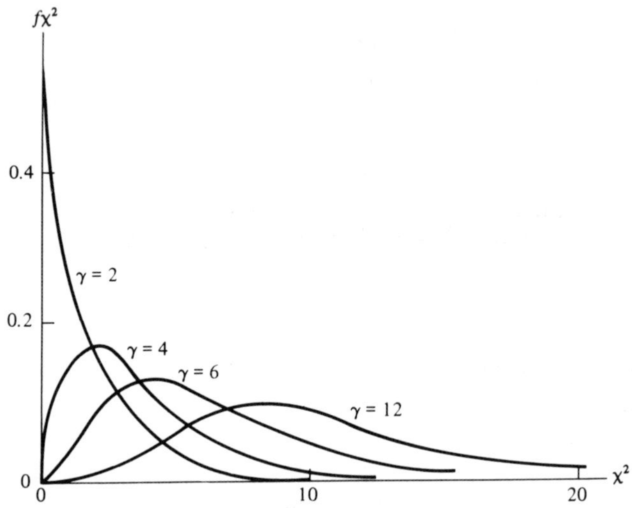

Fig. 4.1. Frequency distribution plots of the chi-square distribution.

and stating what we have said above in symbolic terminology

$$\frac{\Sigma (x_i - \bar{x})^2}{\sigma^2} \tag{4.3}$$

is a random variable having the χ^2 distribution with $\gamma = n - 1$.

Equation (4.3) may be written as

$$\frac{\text{SS}x}{\sigma^2} = \frac{(n - 1)s^2}{\sigma^2} = \chi^2 \tag{4.4}$$

from which the sampling distribution of s^2 is obtained from the distribution of $\chi^2(n - 1)$ so that

$$s^2 = \frac{\sigma^2 \chi^2 (n - 1)}{(n - 1)} \tag{4.5}$$

Equation (4.4) can be used to define confidence intervals. We can do this by first finding the two values of χ^2. For illustrative purposes call them χ_1^2 and χ_2^2, so that

$$p(\chi_1^2 < \chi^2 < \chi_2^2) = \int_{\chi_1^2}^{\chi_2^2} f(\chi^2; \gamma)\, d\chi^2 = 1 - \alpha \tag{4.6}$$

and rearranging using (4.4), we get

$$p\left[\frac{SSx}{\chi_2^2} < \sigma^2 < \frac{SSx}{\chi_1^2}\right] = 1 - \alpha \tag{4.7}$$

For engineering purposes, three uses of the chi-square distribution are of interest: chi-square tests, chi-square goodness of fit, and Bartlett's test for homogeneity of variances.

4.1 CHI-SQUARE TESTS

It has been noted above that chi-square is a measure of variance or deviation (as contrasted with t, which is a measure of difference of means). As such,

$$\chi^2 = \Sigma \frac{(\text{observed frequency} - \text{expected frequency})^2}{\text{expected frequency}}$$

$$= \Sigma \frac{(0 - E)^2}{E} \tag{4.8}$$

The ideal case would be that where there is no deviation; thus χ^2 would be zero. As can be noted, the χ^2 table in Appendix 6, like the t distribution, is dependent on degrees of freedom and α. The table is also one of limiting values. That is, with the null hypothesis of variance, if the calculated χ^2 is less than the tabular value, then we cannot assume any difference in variances.

Example: An automotive manufacturer installs 20% of radio type x and 80% of type y. After 6 months, there have been 2655 failures, 663 of type x and 1989 of type y. Do we have any reason to assume a difference between the two types of radio?

If there is no difference between the radios, we would expect the failure to be in the same ratio as those placed in the automobiles; that is,

$$(0.20)(2655) \text{ of } x = 531$$
$$(0.80)(2655) \text{ of } y = 2124$$

Therefore from (4.8)

$$\chi^2 = \frac{(663 - 531)^2}{531} + \frac{(1989 - 2124)^2}{2124} = 32.9 + 8.6 = 41.5$$

Now assuming a probability of 0.01 with 1 deg of freedom, the tabular χ^2 value is 6.635. Therefore the hypothesis of equality is rejected and we assume that there is a difference.

Many engineering problems are of the type just illustrated. They have but 1 deg of freedom (d.f.). Because χ^2 is a continuous statistical distribution, applying it to 1 d.f. problems introduces a bias which results in variations which are too high. This bias may be compensated for by subtracting 0.5 from the absolute value of each deviation *prior* to squaring it. It must be emphasized that this *correction for continuity* applies only where 1 d.f. is involved. The expression is

$$\chi^2 \text{ corrected} = \Sigma \frac{(\text{observed frequency} - \text{expected frequency} - 0.5)^2}{\text{expected frequency}} \tag{4.9}$$

Example: In inspecting material where 5% defective is the upper specified limit, we find 6 defective units in a sample of 80 units. What do you conclude?

	Observed	Expected	Difference	Difference − 0.5	$\frac{(\text{Difference} - 0.5)^2}{\text{Expected}}$
Unsatisfactory	6	4	2	1.5	0.56
Satisfactory	74	76	2	1.5	0.03
	80	80		Corrected χ^2 =	0.59

This value of 0.59 is compared with the tabular probability value of between 0.5 and 0.3, indicating that there is approximately 0.4 probability of getting a number of rejects as large as that which we obtained when the population actually had only 5% unsatisfactory.

The χ^2 test, as we have seen, can be used to compare some distribution with some other known distribution and determine if there is a difference using the sums of squares techniques. χ^2 is also applied to determine whether a set of data is homogeneous without reference to an outside standard, i.e., distribution.

The data are arranged in a format so that rows and columns represent classes and types. Then the homogeneity of the distributions is tested on the totals. These formats are called *contingency tables*, which can assume any size. The radio example above was an example of a single classification problem. Two-way classification tables test whether the various columns (or rows) have the same proportions in the various categories.

The 2 × 2 case arises when both criteria of classification have two categories. The number of degrees of freedom for the generalized $r \times c$ contingency table is the product of one less than the number of rows by one less than the number of columns. The generalized case with 1 d.f. is

	First Criteria		
Second Criteria	a	b	$a + b$
	c	d	$c + d$
	$a + c$	$b + d$	

In this case the χ^2 statistic is

$$\chi^2 = \frac{(ad - bc - 0.5)^2(a + b + c + d)}{(a + b)(a + c)(b + d)(c + d)} \tag{4.10}$$

Example:

Number of Tire Failures

Test Car	Tire *A*	Tire *B*	
1	18	13	31
2	10	12	22
	28	25	

From (4.10)

$$\chi^2 = \frac{(216 - 130 - 0.5)^2(53)}{(31)(28)(25)(22)} = 0.70$$

$$\therefore \chi^2 < \chi^2_{0.05,1} = 3.84$$

so we accept the hypothesis that we cannot say there is a difference between the categories.

4.2 BARTLETT'S TEST

Somewhat later on we will have use for a test of homogeneity of variances, and since Bartlett's test is a χ^2 test, it will be discussed here. The test compares the difference between the total degrees of freedom multiplied by the logarithm of the pooled estimate of the variance and $\sum_{i=1}^{n}$ (d.f.) (log of estimate of variance), where

n = number of variance compared
k_i = sample size of each sample
s_i^2 = variance of each sample

The number of degrees of freedom associated with the χ^2 value is $n - 1$. The test is

$$\chi^2 = 2.3026 \{(\log \bar{s}^2) \Sigma (k_i - 1) - \Sigma [(k_i - 1) \log s_i^2]\} \quad (4.11)$$

Chi-square, as calculated from (4.11), is biased on the high side and needs to be corrected in accordance with the number of degrees of freedom. The correction factor is

$$c = 1 + \frac{1}{3(n-1)}\left[\Sigma\left(\frac{1}{k_i - 1}\right) - \frac{1}{\Sigma(k_i - 1)}\right] \quad (4.12)$$

and

$$\chi_c^2 \text{ (corrected)} = \chi^2/c$$

Example: Consider the data for a series of determinations on the percent sulfur in 5 west Texas crude oils. Test the variances for homogeneity to see if there is a significant difference among them.

Crude, n	Sample Size, k	Variance Estimate, s^2
1	15	0.68
2	21	0.52
3	24	0.73
4	18	0.84
5	20	0.59

Now, accumulating information for (4.11),

n	$k-1$	SSx	s_i^2	$\log s_i^2$	$(k-1)\log s_i^2$	$1/(k-1)$
1	14	9.52	0.68	0.8327 − 1	11.6550 − 14	0.0715
2	20	10.40	0.52	0.7160 − 1	14.3200 − 20	0.0500
3	23	16.79	0.73	0.8633 − 1	19.8559 − 23	0.0435
4	17	14.28	0.84	0.9243 − 1	15.7131 − 17	0.0589
5	19	11.21	0.59	0.7709 − 1	14.6471 − 19	0.0526
	93	62.20		4.1070 − 5	76.1911 − 93	0.2765

$$\bar{s}^2 = \frac{\Sigma\,\mathrm{SS}x}{\Sigma(k-1)} = \frac{62.20}{93} = 0.6688 \tag{4.13}$$

$$\operatorname{Log} \bar{s}^2 = 0.8254 - 1$$

so substituting in (4.11) and noting that 0.8254 − 1 is the same as

$$\begin{array}{r} \log \bar{s}^2 \\ \hline -1.0000 \\ 0.8254 \\ \hline -0.1746 \end{array} \qquad \text{and similarly} \qquad \begin{array}{r} (k-1)\log s^2 \\ \hline -93.0000 \\ 76.1911 \\ \hline -16.8089 \end{array}$$

$$\begin{aligned} \chi^2 &= 2.3026\,[(-0.1746)(93) - (-16.8089)] \\ &= 2.3026(-16.24 + 16.81) = 1.31 \end{aligned}$$

but from (4.12) we get

$$\begin{aligned} c &= 1 + \frac{1}{3(5-1)}\left(0.2765 - \frac{1}{93}\right) \\ &= 1 + 0.83(0.2765 - 0.01) = 1 + 0.0222 = 1.0222 \end{aligned}$$

so

$$\chi_c^2 = \frac{1.31}{1.022} = 1.28$$

and

$$\chi^2_{0.05,4} = 9.488$$

Therefore there is less than 1 chance in 20 that these variance estimates are from samples from different populations.

If the size of each of the samples is the same, then Eq. (4.11), (4.12), and (4.13) simplify to

$$\chi^2 = 2.3026(k-1)(n \log \bar{s}^2 - \Sigma \log s_i^2) \tag{4.14}$$

$$c = 1 + \frac{n+1}{3n(k-1)} \tag{4.15}$$

and

$$\bar{s}^2 = \frac{\Sigma s^2}{n} \tag{4.16}$$

4.3 GOODNESS OF FIT TEST

Throughout the whole statistical spectrum, we have a common problem: How far can our observed distribution vary from the theoretical one and still let us assume that the observed is simply a sample of the theoretical? Frequently we want a yes or no answer. The chi-square goodness of fit test can give us this answer (with an α probability of being wrong).

It should be emphasized that this test can be applied to *any* distribution, although we will use the normal distribution to illustrate the method. In Chapter 3 we learned how to calculate theoretical frequencies for the binomial, Poisson, and other distributions. In addition to learning the goodness of fit test here, we will see how to calculate the theoretical frequencies for the normal curve. The columns in Table 4.1 will be numbered for easy reference. The mean of the data is 0.5163 in. with a standard deviation of 0.0052 in. There are 244 samples. Column (1) results from dividing the distribution into classes.

TABLE 4.1 Cylinder Diameters, Inches

Mid-value, in. (1)	Observed Frequency (2)	Class Boundary (3)	*z* for Class Boundary (4)	Area Below *z* (5)	Area Within Class (6)	*n* [column (6)] (7)
0.500	10				0.0040	0.98
		0.5025	−2.65	0.0040		
0.505	21				0.0415	10.13
		0.5075	−1.69	0.0455		
0.510	38				0.1872	45.68
		0.5125	−0.73	0.2327		
0.515	95				0.3583	87.42
		0.5175	0.23	0.5910		
0.520	46				0.2920	71.24
		0.5225	1.19	0.8830		
0.525	26				0.1012	24.69
		0.5275	2.15	0.9842		
0.530	8				0.0158	3.86
	244				1.0000	244.00

Column (2) is the tally within the chosen classes. Column (3) is simply the boundary between classes. Column (4) is the z value corresponding to each class boundary. The general expression *for this problem* is

$$z = \frac{x - \bar{x}}{\sigma} = \frac{x - 0.5163}{0.0052} = 192.3076x - 99.2884 \quad (4.17)$$

so that if the first x from column (3) is substituted in (4.17), we get

$$\begin{aligned} z &= 192.3076(0.5025) - 99.2884 \\ &= 96.6346 - 99.2884 = -2.654, \text{etc.} \end{aligned}$$

until column (4) is complete. Column (5) is the area from $-\infty$ up to the class boundary, i.e., it is cumulative. It is obtained from a table of z values as shown in Appendix 4. The negative values can be read directly from the table, but remember that all positive values must be subtracted from 1.0000. Column (6) is the difference between the values in column (5) or, actually, the area between class boundaries. Column (7) is the number of cylinders in each class, found by multiplying the

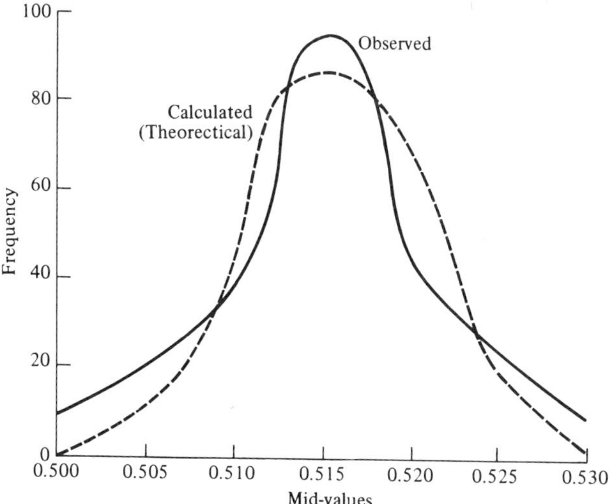

Fig. 4.2 Frequency distribution plot of observed and theoretical values for the data shown in Table 4.1.

total of 244 times the percent of area within each class that is under the normal curve. Figure 4.2 shows the observed and calculated frequencies in Table 4.1.

Up to this point we have merely been graduating the curve. Now let us apply the χ^2 test. From (4.8), we set up Table 4.2.

TABLE 4.2 Chi-Square Goodness of Fit Test of Data in Table 4.1

Mid-value	Observed Frequency, f_o	Calculated Frequency, f_c	$f_o - f_c$	$(f_o - f_c)^2/f_c$
0.500	10	0.98	9.02	83.02
0.505	21	10.13	10.87	10.97
0.510	38	45.68	7.68	1.29
0.515	95	87.42	7.58	0.66
0.520	46	71.24	25.24	8.94
0.525	26	24.69	1.31	0.07
0.530	8	3.86	4.14	4.44
				109.39

$$\chi^2 = \Sigma \frac{(f_o - f_c)^2}{f_c} = 109.39$$

In the case of the test for goodness of fit, the degrees of freedom are the number of classes minus 3. One degree of freedom is lost because the sum of the calculated frequencies was made equal to the observed total frequency and we lose two more because the two parameters $\bar{x}$ and σ were used in graduating the distribution.

Therefore the above χ^2 is compared with (assuming α = 0.01)

$$\chi^2_{0.01,4} = 13.28$$

Since the calculated χ^2 is greater than the tabular χ^2, we reject the hypothesis of equality and assume, within the 0.01 chance of error, that the observed distribution is not a normal one. Note that the first class deviation is the one which causes the decision. The larger the chi-square value, the poorer the fit according to our criterion.

To illustrate the goodness of fit test using the binomial distribution, let us repeat the data from Table 3.1 from Chapter 3 in columns (1) and (3) in Table 4.3. Assume 100 items. $\chi^2_{0.01,7}$ = 18.48, which indicates that the observed distribution

TABLE 4.3

Number of Defects in Sample, x (1)	Observed Frequency, f_o (2)	Calculated Frequency, f_c (3)	$f_o - f_c$ (4)	$(f_o - f_c)^2/f_c$ (5)
0	15	13.26	1.74	0.23
1	24	27.06	3.06	0.35
2	28	27.34	0.66	0.02
3	20	18.23	1.77	0.17
4	8	9.02	1.02	0.12
5	4	3.54	0.46	0.06
6	1	1.14	0.14	0.02
7	0	0.31	0.31	0.31
8	0	0.07	0.07	0.07
9	0	0.02	0.02	0.02
			χ^2 =	1.37

can be assumed within the 0.01 chance of error to be a binomial.

Note the value 0.31. This one value accounts for approximately 25% of the error, although the difference between f_o and f_c is very small, especially in view of the larger figures in the table. This phenomenon requires the use of a process called *lumping*. The rule for lumping dictates that adjacent classes with less than a frequency of 5.0 should be combined. Of course, a degree of freedom is lost each time you lump. If we were to analyze the data in Table 4.3 properly, it would be as shown in Table 4.4. $\chi^2_{0.01,3} = 11.35$. Thus it can be seen that the right tail of the distribution contributed 0.48 to the chi-square value in Table 4.3 and contributed only 0.001 in Table 4.4.

TABLE 4.4

Number of Defects in Sample, x	Observed Frequency, f_o	Calculated Frequency, f_c	$f_o - f_c$	$(f_o - f_c)^2/f_c$
0	15	13.26	1.74	0.23
1	24	27.06	3.06	0.35
2	28	27.34	0.66	0.02
3	20	18.23	1.77	0.17
4	8	9.02	1.02	0.12
5	5	5.08	0.08	0.001
			χ^2 =	0.891

5
Analysis of Variance

Analysis of variance is a computational process for separating a total variance into its component parts for the purposes of statistical inference. The parts are then arranged so that structured similarities and dissimilarities can be gleaned. Statistical tests can then be used to test the resulting hypothesis.

Circumstances often dictate that we design a test in such a manner that several variables can be studied simultaneously. If we wanted to test differences among six means using the t tests on each pair of means, there would be $\binom{6}{2}$ or 15 t values to compute. If we decided to run all the t tests, we run the chance of some of the values exceeding t when, in fact, they do not.

Analysis of variance (ANOVA) is our first step toward model construction. As in hypothesis testing, we must hypothesize a model form. In the consideration of ANOVA problems, we have varying limitations on how we arrange the data so that they will yield the information we want. Perhaps the most common form or arrangement is the completely randomized design where the treatments are assigned completely at random to the experimental units in order to minimize bias.

The basic assumption for a completely randomized design having one observation per experimental unit is that observations can be represented by the model

$$x_{ij} = \mu + \tau_i + \epsilon_{ij} \tag{5.1}$$

where x_{ij} = any observation
μ = true mean effect
τ_i = true effect of the ith treatment
ϵ_{ij} = commonly called the error term

μ is a constant since we possess the data. ϵ_{ij} is normally and independently distributed with a mean of zero and a variance of σ. The ϵ term includes statistical misfit of data to model, nonadditivity components of variables, i.e., treatments, and other extraneous factors.

Two statements can be made about τ. The first can be represented by

$$\sum_{i=1}^{t} \tau_i = 0 \qquad \text{or} \qquad H_0\colon \tau_i = 0 \qquad (i = 1, \ldots, t) \tag{5.2}$$

This statement, in effect, means that we are concerned only with the data under consideration in this experiment and that we do not wish to extrapolate our findings to some larger population. The model in (5.2) is known as *model I* or the fixed effects ANOVA model.

The other statement considers the τ_i's as merely samples from a normally and independently distributed population with a mean of zero and a variance of σ. This is *model II* or the random effects ANOVA model. Here the engineer is concerned with inferences about some larger population of which his data are but a sample. We must be concerned with determining which model we shall use because the inferences and analysis procedures are different.

And speaking of inferences, just what do we hope to learn from the ANOVA procedure? (1) hypotheses about the effects of treatments, (2) estimation of the magnitude of components of variance, and (3) estimation of the *mean* effects of treatments.

Prior to the discussion of any ANOVA models, certain assumptions must be made about the observations.

The first of these is *additivity*. In any experiment there are two possible differences between samples: that produced by varying conditions (we call these conditions treatments) and that produced by sampling differences. The treatment effects can be separated into two categories: those which are additive and constant and those which are not. The property of "additive and constant" is that which produces the same increment of response for every sample which receives the same treatment, given that the samples are the same.

There are three primary causes of nonadditivity. The true effects may be multiplicative; i.e., we may have an effect of x_1 times x_2 in addition to $x_1 + x_2$. A certain yield might be realized when pressure x_1 and temperature x_2 have an interacting effect. Hence the x_1x_2 term is called an *interaction* or *cross-product* term. The second cause lies in the fact that the interactions may exist but may be neglected by the engineer. The third cause could result from deranged or spurious data points.

The second assumption is that of *homogeneity of variances.* Bartlett's test, discussed in Chapter 4, is the appropriate test for determining whether or not this homogeneity exists. We are determining whether or not the variances have come from the same universe.

The third assumption is that of *normality*, which can be verified using the chi-square test of goodness of fit. An alternative is the Kolmogorov-Smirnov test, which is not discussed here but can be found in many statistical reference books.

The fourth assumption is that of *independence.* Independence denotes uncorrelated errors. Randomization is the best device to prevent bias. The control chart approach depicting the pattern of scatter of sample parameters is one device helpful in this area.

The question that is in order at this point is how much deviation can exist in these underlying prerequisite assumptions for our results to be valid? Generally, the consequences are not serious if the basic assumptions are not strictly adhered to. Fortunately, the ANOVA method is rather strong and will yield worthwhile information.

Is there anything that can be done if we have done our work with care and still do not get meaningful results. First, the data may be unrelated to the extent that we can draw no conclusions to the possible relationships. We must always be ready to admit this possible outcome. Second, we may investigate some of the many transforms that are available. *Transformations* are used for two purposes. First, we might want to normalize the error distribution, and, second, we might want to create a scale where treatment effects are additive. If the relationship is multiplicative rather than additive, then some sort of log transform might be in order. For instance,

$$y = a_1 x_1 w^2$$

can be written

$$\log y = \log a + \log x + 2 \log w$$

and we are back to our additive situation. In addition to using the logarithmic transformation when true effects are multiplicative, we would use it when the standard deviation is proportional to the mean.

The square-root transform, familiar to all engineers, would be used when the variance is proportional to the mean. The reciprocal transformation is used when the standard deviation is proportional to the square of the mean. Fortunately there are screening programs available for all digital computing equipment. These programs consider an array of data and give the engineer information which enables him to choose the transform which best fits the data.

5.1 REEXAMINATION OF VARIANCE

Now that we have some experience in working with variance, let us look at it again. We should now have the feeling that a study of variance is almost synonymous with the field of statistics itself. If no variance existed, we would have little or no use for statistics. The understanding and prediction of variance is the *raison d'être* of statistics.

The variance of a sum (or difference) of independently varying functions is equal to the sum of the individual variances. If

$$y = x_1 + x_2 + \cdots + x_n$$

or

$$y = x_1 - x_2 - \cdots - x_n \qquad (5.3)$$

then

$$\sigma^2(y) = \sigma^2(x_1) + \sigma^2(x_2) + \cdots + \sigma^2(x_n)$$

or

$$\sigma^2(y) = \sigma^2(x_1) - \sigma^2(x_2) - \cdots - \sigma^2(x_n) \qquad (5.4)$$

assuming that the x's vary independently. In Chapter 3 we saw how we could attribute confidence ranges to variables using the expression

$$x \pm t_{\alpha,\text{d.f.}} s(x) \qquad (5.5)$$

If we have an estimate of the variance, we can calculate the *precision limits* (p.l.) of x. The values $\pm t_{\alpha,\text{d.f}} s(x)$ are referred to as the precision limits.

Example: We have a hydrocarbon stream made up of 2000 gpm of C_2, 5000 gpm of C_3, and 1,000 gpm of C_4. If the standard deviation of each measurement is 6% of the individual streams, calculate the 90% p.l. for each cut and the 90% p.l. and confidence range for the total flow. A 90% precision limit is equivalent to 1.645σ since 1.64 is the 0.10 value of t at ∞ d.f. From expression (5.4) we can formulate the following:

Stream	Flow, gpm	σ	σ^2	90% p.l.
C_2	2000	120	14,400	±196.8
C_3	5000	300	90,000	±492.0
C_4	1000	60	3,600	±98.4
	8000		108,000	±787.2

The total of the p.l.'s is 787.2 for a stream of 8000 gpm. But by considering (5.4) we get

$$\text{Total flow } \sigma^2 = 108{,}000$$

Therefore

$$\sigma = 328.6$$

and

$$\text{Total flow p.l.} = \pm 1.64\,(328.6) = \pm 538.9$$

and therefore the confidence range would be, from (5.5),

$$8000 \pm 538.9$$

If instead of a sum or difference we had the variance of a function, then it would be handled as illustrated below:

Function	Variance
$y = kx$, where k = constant	$\sigma^2(y) = k^2\sigma^2(x)$
$y = x_1 x_2$	$\sigma^2(y) = \bar{x}_2^2\sigma(x_1) + \bar{x}_1^2\sigma(x_2)$
$y = \dfrac{x_1}{x_2}$	$\sigma^2(y) = \dfrac{\bar{x}_2^2\sigma(x_1) + \bar{x}_1^2\sigma(x_2)}{\bar{x}_2^4}$

In cases where $f(x_1, x_2, \ldots, x_n)$ cannot be assumed to be distributed normally, *Tchebycheff's inequality* can be resorted to. This theorem states that for any distribution with finite mean and variance, at least $(1 - 1/z^2)$ 100% of the values are distributed in $a \pm z\sigma$ range around the average μ. This theorem is an approximation, and the sample size should be rather large.

5.2 F TESTS FOR VARIANCE

We have investigated situations where we wanted to find the distribution of a sample variance. We now turn to the case of the distribution of the ratio of variances of two independent random samples. This test, known as the F test in honor of R. A. Fisher, the famous statistician, is widely used and most important in regression-correlation analysis and analysis of variance. We want to test whether two samples come from populations having equal variances.

If s_1^2 and s_2^2 are variances of independent random samples of size n_1 and n_2, respectively, from populations which are normally distributed and have the same variance, then

$$F = \frac{s_1^2}{s_2^2} \tag{5.6}$$

is a value of a random variable having a distribution known as the F distribution with the parameters $\nu_1 = n_1 - 1$ and $\nu_2 = n_2 - 1$, which are the degrees of freedom for the numerator and denominator variances, respectively.

Like the t and χ^2 tables, the F tables are ones of limiting values. That is, if the calculated F is larger than the corresponding tabular F, the distributions are assumed to be different with an α chance of error. Appendix 7 contains values of $F_{0.01}$ and $F_{0.05}$ for various combinations of degrees of freedom. Note that these are right-tail areas. The tables can be used for the similar left tails using the model

$$F_{1-\alpha;\nu_1,\nu_2} = \frac{1}{F_{\alpha;\nu_2,\nu_1}} \tag{5.7}$$

The F test can be applied to a ratio of variances or to any ratio of values which are related to variances, such as differences in sums of squares, differences in correlation coefficients (to be covered in Chapter 6), etc.

5.3 SINGLE CLASSIFICATION MODELS

Although we used a t test to determine the difference between two means, we will use an elementary ANOVA model to compare several means in terms of the pooled variance of the sample measurements.

If we encountered the situation where we had a process where we obtained yields at different levels of pressure, feed stock, temperature, and production units, we would want to separate the effects of these factors if there were, in fact, a difference. The problem is further complicated by the presence of a measurement error as well as the variation due to the physical characteristics of the process.

Each type of ANOVA model possesses a unique format for arranging the data for analysis. The format for the single classification model is shown in Table 5.1.

TABLE 5.1

Groups	A_1	A_2	$\cdots$	A_j	$\cdots$	A_n
Measurements	x_{11}	x_{12}	$\cdots$	x_{1j}	$\cdots$	x_{1n}
	x_{21}	x_{22}	$\cdots$	x_{2j}	$\cdots$	x_{2n}
	$\vdots$	$\vdots$		$\vdots$		$\vdots$
	$x_{k_1 1}$	$x_{k_2 2}$	$\cdots$	$x_{k_j j}$	$\cdots$	$x_{k_n n}$
Number of observations	k_1	k_2	$\cdots$	k_j	$\cdots$	k_n

Total number of observations: $N = \sum_{i=1}^{n} k_j$

Over-all mean: $\bar{x} = \sum_{i=1}^{k} \sum_{j=1}^{n} \frac{\bar{x}_{ij}}{\mathrm{N}} = \Sigma x / N.$

The sums of squares can be apportioned so that we determine whether or not a difference exists between and within rows and columns. You will recognize that a difference between columns is a difference between groups, whereas a difference *within* a column is a measurement or error difference. The repeated measurements under each group are called a replicate (repetition), and, by doing this, we can obtain a variance while all physical conditions remain the same. This variance or error can be attributed to lack of precision in measuring technique or to several other sources of error which will be discussed later.

The test is to determine whether any of the groups (A's) has a significant effect on any variation that may be found in the x's. We will compare the variance estimates within columns (replicates) and the variance estimates between columns (sample means). If this ratio exceeds the tabular $F_{\alpha,(n-1,N-n)}$, we reject the hypothesis that two mean squares come from the same distribution; therefore we accept the fact that group A contributes the difference.

To obtain the variance estimates, we will divide the sums of squares of deviations from the mean by the degrees of freedom. The model for the total sum of squares with $N - 1$ degrees of freedom is

$$\sum_{i=1}^{n}\sum_{j=1}^{k}(x_{ij} - \bar{x})^2 = \Sigma x^2 - \frac{(\Sigma x)^2}{N} \tag{5.8}$$

The between-group sum of squares, with $n - 1$ degrees of freedom, is

$$\sum_{i=1}^{n} k_i(\bar{x}_i - \bar{x})^2 \tag{5.9}$$

and we can obtain the within-group sum of squares, with $\Sigma (k_i - 1)$ degrees of freedom, by difference.

Example: The data in Table 5.2 show four replicate measurements on the octane number of a fuel tested on five CFR machines. We are asked to determine if there is a difference between the machines.

TABLE 5.2

CFR	1	2	3	4	5
Measurement 1	85.2	87.2	86.9	83.0	85.0
Measurement 2	85.7	86.4	87.0	83.8	86.1
Measurement 3	84.9	86.6	86.4	83.9	86.5
Measurement 4	85.5	86.8	86.5	83.8	85.5
CFR mean	85.3	86.8	86.7	83.7	85.9

The figures in Table 5.2 are somewhat cumbersome to deal with. In most cases some form of *coding* can be made use of. Here if we subtract 80 and multiply 10, we will obtain the data in Table 5.3, and since we are dealing with variances and ratios of variance, nothing is lost.

TABLE 5.3

CFR	1	2	3	4	5
Measurement 1	52	72	69	30	50
Measurement 2	57	64	70	38	61
Measurement 3	49	66	64	39	65
Measurement 4	55	68	65	38	55
CFR mean	53	68	67	37	59
Σ	213	270	268	145	231

Now we must formulate the ANOVA format in general and then as it specifically pertains to the case being considered. Table 5.4 is the general form.

Now, for the case under consideration,

$$\Sigma X = 52 + 57 + 49 + \cdots + 65 + 55 = 1127$$

$$\Sigma X^2 = (52)^2 + (57)^2 + (49)^2 + \cdots + (65)^2 + (55)^2 = 66{,}401$$

$$\Sigma (A_i)^2 = 213^2 + 270^2 + 268^2 + 145^2 + 231^2 = 264{,}479$$

$$(\Sigma X)^2 = 1127^2 = 1{,}270{,}129$$

$$N = 20$$

$$n = 5$$

and

$$k = 4$$

So the specific quantities required for Table 5.4 are

$$\frac{(\Sigma X)^2}{N} = \frac{1{,}270{,}129}{20} = 63{,}506.45$$

and

$$\sum \frac{(A_i)^2}{k_i} = \frac{(213)^2}{4} + \frac{(270)^2}{4} + \frac{(268)^2}{4} + \frac{(145)^2}{4} + \frac{(231)^2}{4}$$

$$= \frac{264{,}479}{4} = 66{,}119.75$$

TABLE 5.4

Source	Sum of squares (SS)	Degrees of Freedom (d.f.)	Mean Square (MS)
Between groups	$\sum \frac{(A_i)^2}{k_i} - \frac{(\Sigma X)^2}{N}$	$n - 1$	Between groups SS/$(n - 1)$
Within groups	Difference	Difference	Difference SS/Difference d.f.
Total	$\Sigma X^2 - (\Sigma X)^2/N$	$N - 1$	—

TABLE 5.5

Source	SS	d.f.	MS
Between groups	$\sum \frac{(A_i)^2}{k_i} - \frac{(\Sigma X)^2}{N} =$ $66{,}119.75 - 63{,}506.45 = 2613.30$	$n - 1 = 4$	$\frac{2613.30}{4} = 653.32$
Within groups	By difference, $2894.55 - 2613.30 = 281.25$	$19 - 4 = 15$	$\frac{281.25}{15} = 18.75$
Total	$\Sigma X^2 - (\Sigma X)^2/N =$ $66{,}401 - 63{,}506.45 = 2894.55$	$N - 1 = 19$	

The F test for the data in Table 5.5 is $F = 653.32/18.75 = 34.84$, and the tabular F for $\alpha = 0.05$ and 4 and 15 d.f., from Appendix 7, is 3.06. Since the calculated F is greater than the tabular F, we reject the hypothesis of no difference and conclude that there is evidence of a factor between CFR engines which causes more variation in the data, with an 0.05 probability of error, than we can account for by the variability within groups.

Previously we discussed the difference between model I and model II. The above discussion pertains to model I, and for the single classification case, we must discuss model II. It will be remembered that model II involves variation in means and is applicable when the random factor is defined by a distribution where our data are only a sample. We will define the within-group source as the residual, or error, sum of squares, since it represents variation when everything else is, hopefully, constant.

We designate the residual mean square, which is the combination of error estimate and unaccessible error, as σ^2. The between-group mean square is the estimated error variance to which is added the variance attributable to the factor in question (if any) multiplied by k, k being the number of samples in the group, i.e., $\sigma^2 + k\sigma^2(A)$. Therefore, for the single classification case, model II, we have Table 5.6.

TABLE 5.6

Source	SS	d.f.	MS	Variance Estimated
Between groups	$\sum \frac{(A_i)^2}{k_i} - \frac{(\Sigma X)^2}{N}$	$n - 1$	SS/d.f.	$\sigma^2 + k\sigma^2(A)$
Error	Difference	Difference	SS/d.f.	σ^2
Total	$\Sigma X^2 - (\Sigma X)^2/N$	$N - 1$	—	—

In the case under discussion the F test would be

$$F = \frac{\sigma^2 + k\sigma^2(A)}{\sigma^2} \tag{5.10}$$

Now that we are familiar with the sums of squares and mean squares calculation, we will denote these quantities using the following shorthand notations:

$$\begin{aligned} SS_b &= \text{between-group sum of square} \\ MS_b &= \text{between-group mean square} \\ SS_e &= \text{error (within groups) sum of square} \\ MS_e &= \text{error mean square} \\ SS_{to} &= \text{total sum of square} \end{aligned}$$

The subscripts attached to SS and MS notations will be defined as they are encountered. For our purposes we will use the "variance-estimated" column to realize what F tests we should make in model II. The actual test will be a ratio of mean squares as in model I. In formulating F tests in model II, we are relatively free to formulate these ratios with the restriction that the numerator must have only one more term than the denominator, σ^2, $k\sigma^2(A)$, etc., being terms.

5.4 SINGLE CLASSIFICATION WITH SUBGROUPINGS (NESTING)

Frequently we have a hierarchy of variables such that subgroups are important. We might be interested in one pipe still run at two different vacuums with three different crudes run at each vacuum. Diagrammatically, see Figure 5.1. The numbers

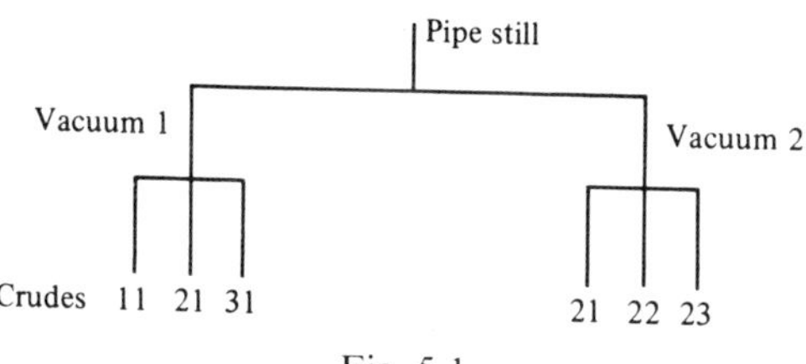

Fig. 5.1.

11, 21, etc., indicate the first crude at the first vacuum, the second crude at the first vacuum, etc. If more than one reading were taken for each crude at each vacuum level, then we would have replicate measurements which would enable us to get an error (within-group) estimate.

Table 5.7 gives the generalized ANOVA table for the single classification with subgroups. But before we discuss the table,

TABLE 5.7

Source	SS	d.f.	MS	Variance Estimates
A	$\sum \frac{(A\text{ totals})^2}{kbcd} - \frac{(\Sigma X)^2}{N}$	$a - 1$	$SS_{A/(a-1)}$	$\sigma^2 + k\sigma^2 D + kd\sigma^2(C) + kcd\sigma^2(B) + kcbd\sigma^2(A)$
$B(A)$	$\sum \frac{(B\text{ totals})^2}{kcd} - \sum \frac{(A\text{ totals})^2}{kbcd}$	$a(b - 1)$	$SS_{B(A)/[a(b-1)]}$	$\sigma^2 + k\sigma^2(D) + kd\sigma^2(C) + kcd\sigma^2(B)$
$C(AB)$	$\sum \frac{(C\text{ totals})^2}{kd} - \sum \frac{(B\text{ totals})^2}{kcd}$	$ab(c - 1)$	$SS_{C(AB)/[ab(c-1)]}$	$\sigma^2 + k\sigma^2(D) + kd\sigma^2(C)$
$D(ABC)$	$\sum \frac{(D\text{ totals})}{k} - \sum \frac{(C\text{ totals})^2}{kd}$	$abc(d - 1)$	$SS_{D(ABC)}/[abc(d - 1)]$	$\sigma^2 + k\sigma^2(D)$
Error	$\sum X^2 - \sum \frac{(D\text{ totals})^2}{k}$	$abcd(k - 1)$	$SS_E/[abcd(k - 1)]$	σ^2
Total	$\Sigma X^2 - (\Sigma X)^2/N$	$N - 1$		

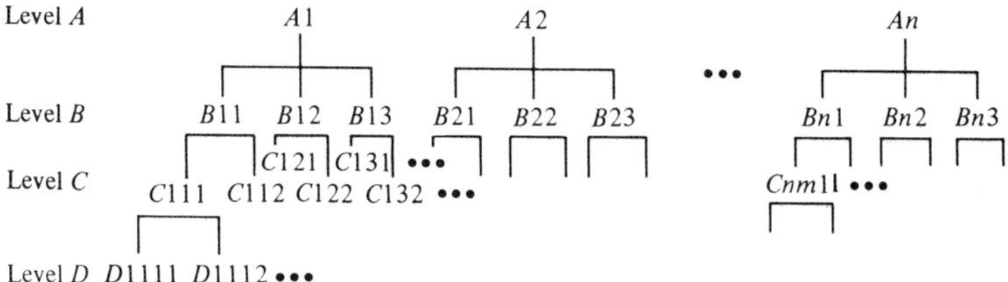

Fig. 5.2.

consider the hierarchy in Figure 5.2. $A1$ indicates the first or top level; $B11$ indicates the first sample on the B level under $A1$; $B12$ indicates the first sample on the B level under $A2$; etc. Similarly, $C111$ is the first sample under $B1$ and $A1$, etc. It should be emphasized that, in Figure 5.2, all numbers are indexing constants, not sample measurements. Now it is necessary to define the nomenclature to be used in Table 5.7. From Figure 5.2 the first level is factor A with the number a samples, under A is the subgroup factor B, denoted $B(A)$ with b samples, under B is the subgroup factor C, denoted $C(AB)$ with c samples, under C is the subgroup factor D, denoted $D(ABC)$ with d samples; and under D there are k replicates under each D. Therefore the total number of samples, N, is $a \cdot b \cdot c \cdot d \cdot k$. Note that if we wanted to determine whether a difference existed between A and $B(A)$, we could formulate an F test since the variance estimated for A has one and only one more term than $B(A)$. For this reason we could not test A versus $C(AB)$ because we could not separate any discovered difference between the subgroups of B and A.

Example: The data in Table 5.8 show the results of a situation of three crude distillation units, A's, two different crude oils, B's, two different vacuums, C's, with two measurements of each result. Therefore from Fig. 5.2 and Table 5.7 we would have $abck$ total measurements, or $3 \cdot 2 \cdot 2 \cdot 2 = 24$. We are using the same indexing nomenclature as in Figure 5.2.

$$\sum^{24} X = 21 + 24 + 23 + \cdots + 29 + 27 = 554$$

$$\sum^{24} X^2 = 21^2 + 24^2 + 23^2 + \cdots + 29^2 + 27^2 = 13{,}054$$

TABLE 5.8

A Units	$B(A)$ 2 Crudes	$C(AB)$ 2 Vacuums	k 2 Measurements	Totals
1	11	111	21, 24	
		112	23, 25	93
	12	121	20, 22	
		122	24, 25	91
2	21	211	20, 19	
		212	21, 18	
	22	221	19, 20	
		222	21, 19	
3	31	311	22, 26	
		312	27, 25	
	32	321	29, 28	57
		322	29, 27	56

$$\sum (C \text{ totals})^2 = 45^2 + 48^2 + 42^2 + \cdots + 57^2 + 56^2 = 26{,}050$$

$$\sum^{6} (B \text{ totals})^2 = 93^2 + 91^2 + 78^2 + 79^2 + 100^2 + 113^2 = 52{,}024$$

$$\sum^{3} (A \text{ totals})^2 = 184^2 + 157^2 + 213^2 = 103{,}874$$

Now that these combinations have been accumulated we must reduce the totals to a comparable basis by dividing them by the number of measurements which went into each quantity to obtain the totals. In accumulating the C totals, there were $k = 2$ replicates, so

$$\frac{\Sigma\,(C\text{ totals})^2}{k} = \frac{26{,}050}{2} = 13{,}025$$

In accumulating the B totals, there were $kc = 2 \cdot 2 = 4$ measurements $(21 + 24 + 23 + 25 = 93)$, so

$$\frac{\Sigma\,(B\text{ totals})^2}{kc} = \frac{52{,}024}{4} = 13{,}006$$

In accumulating the A totals, there were $kcb = 2 \cdot 2 \cdot 2 = 8$ measurements, so

$$\frac{\Sigma\,(A\text{ totals})^2}{kbc} = \frac{103{,}874}{8} = 12{,}984.25$$

Note the similarity among the above quantities. The correction term is

$$\frac{(\Sigma\,X)^2}{N} = \frac{554^2}{24} = \frac{306{,}916}{24} = 12{,}788.17$$

Substituting the values in the generalized ANOVA tableau (Table 5.7) for this situation, we get Table 5.9. Now, what analysis should we apply to this problem. As has been said, the interpretation that we place on the situation would dictate whether we classify it as model I or model II.

TABLE 5.9

Source	SS	d.f.	MS
A, units	12,984.25 − 12,788.17 = 196.08	2	98.04
$B(A)$, crude within units	13,006 − 12,984.25 = 21.75	3	7.25
$C(AB)$, vacuum within crudes and units	13,025 − 13,006 = 19.00	6	3.17
Error, replicates	13,054 − 13,025 = 29.00	12	2.42
Total	13,054 − 12,788.17 = 265.83	23	

We can subject the C's to the following F test to determine whether that level is significant when evaluated against the error mean square:

$$F = 3.17/2.42 = 1.31 \qquad \text{and } F_{0.05;6,12} = 3.00$$

Therefore, we could conclude (with an α of 0.05) that there is no statistical difference.

To determine whether a difference can be attributed to the units when considered against the crude types, we formulate

$$F = 98.04/7.25 = 13.52 \qquad \text{and} \qquad F_{0.05;2,3} = 9.55$$

Therefore we would attribute a difference here and conclude that there is a difference between units. Suppose we had chosen an α of 0.01. Then $F_{0.01;2,3} = 30.82$. What would this have told us? It says that the test cannot say there is a difference with the small chance of error which is specified. Also notice that the tabular F's get smaller as the degrees of freedom increase (down and to the right on the tables in Appendix 7).

5.5 TWO-FACTOR ANOVA

In Table 5.1, it was noted that there were n groups (A_n) with k replications in each group. There was only one classification and that was the difference between columns. The difference between rows had no meaning. Now, in the two-factor case, both of these differences are meaningful. For purposes of formulating the generalized format for data arrangement for the two-factor ANOVA, we will call column factors C and row factors R; X_{ij} will be the measured value for the intersection of

TABLE 5.10

Column factor:	C_1	C_2	C_3	$\cdots$	C_n
Row factor:					
R_1	X_{11}	X_{12}	X_{13}	$\cdots$	X_{1n}
R_2	X_{21}	X_{22}	X_{23}	$\cdots$	X_{2n}
$\vdots$	$\vdots$	$\vdots$	$\vdots$		$\vdots$
R_m	X_{m1}	X_{m2}	X_{m3}	$\cdots$	X_{mn}

the ith row and jth column. Note that there is no replication in Table 5.10. In a case such as this, where do we get an error term? We obtain the error variance by the difference between the total sum of squares and sum of squares due to the row and column factors.

Example: The following coded data (Table 5.11) depict the effect two different heat lamp drying times have on five different types of paints. Assume a model I and test at the 0.01-level situation.

TABLE 5.11

Drying Time	Paint Type					
min	1	2	3	4	5	Σ
3	3.23	3.56	3.84	3.62	4.05	18.30
5	4.12	4.62	5.06	4.73	4.92	23.45
Σ	7.35	8.18	8.90	8.35	8.94	41.75

$$
\begin{aligned}
c &= 5 \\
r &= 2 \\
N &= 10 \\
\Sigma X &= 41.75 \\
\Sigma X^2 &= 177.86 \\
\Sigma C^2 &= 7.35^2 + 8.18^2 + 8.90^2 + 8.35^2 + 8.97^2 = 350.33 \\
\Sigma r^2 &= 18.30^2 + 23.45^2 = 884.79 \\
(\Sigma X)^2/N &= 41.75^2/10 = 174.31 \\
\Sigma C^2/r &= 350.33/2 = 175.16 \\
\Sigma r^2/c &= 884.79/5 = 176.96
\end{aligned}
$$

TABLE 5.12

Source	SS	d.f.	MS
Paint type	175.16 − 174.31 = 0.85	4	0.21
Drying time	176.96 − 174.31 = 2.65	1	2.65
Error	(By difference) = 0.05	4	0.01
Total	177.86 − 174.31 = 3.55	9	

Since this is a model I case, the F tests are

$$F = 0.21/0.01 = 21; \qquad F_{0.01;4,4} = 15.98$$

and

$$F = 2.65/0.01 = 265; \qquad F_{0.01;1,4} = 21.20$$

Therefore, since the calculated F is, in both cases, greater than the tabular F, we reject the hypothesis of equality and conclude that the effects due both to paint type and drying time are significant (with an 0.01 chance of error).

5.6 TWO-FACTOR ANOVA WITH REPLICATION

Two-factor ANOVA with replication is an important phase in that we begin to consider *interactions*, or *cross-product* terms, as they are sometimes called. Up to this point we assumed that any effect, as indicated by the sums of square total, not attributed to the factors that we were testing automatically reverted to the error term. Actually, the error term is made up of a number of items among which are nonfit of data, measurement inaccuracies, irrelevant factors, and interactions.

We will now consider the methodology whereby we account for the interaction sums of square, thereby identifying some of the variation which was previously in the error (thus unknown or unassignable) term. The ideal situation is that which enables us to explain *all* the variation. If we have level P for pressure and T for temperature, there may be an effect due to P alone, to T alone, *and* to the effect of PT, which is the interaction term. A plastic monomer may not yield a satisfactory product in the presence of a certain range of pressures alone or in the presence of a certain range of temperatures alone but may do so when P and T are both combining to create a satisfactory environment. Thus the PT interaction might explain the yield rather than the additive combination of $P + T$. Some time ago we looked at additive effects such as $P + T$. Multiplicative effects are interaction terms such as PT. In model (5.1) we stated that observations could be represented by

$$X_{ij} = \mu + \tau_i + \epsilon_{ij}$$

and we identified τ_i as the true effect of the ith treatment. If we assumed the pressure-temperature situation without interaction, our model would become

$$X_{ij} = \mu + P + T + \epsilon_{ij} \tag{5.11}$$

and with interaction we would get

$$X_{ij} = \mu + P + T + PT + \epsilon_{ij} \tag{5.12}$$

Since each term in (5.11) and (5.12) is represented by some sum of squares quantity, the error term ϵ_{ij} in (5.11) includes the SS_{PT} quantity, whereas the ϵ_{ij} term in (5.12) is reduced by that quantity. This concept of apportioning of sum of squares is an important one and should be completely understood before leaving this section. We will now expand Table 5.10, which shows the generalized data arrangement for the two-factor ANOVA classification, into one which shows the two-factor with replication case. The expanded table is Table 5.13. It shows the arrangement for only two replications ($k = 2$), but it can easily be seen that it can be expanded into k replicates if need be.

TABLE 5.13

Column factor:	C_1	C_2	C_3	$\cdots$	C_n
Row factor					
R_1	X_{111}	X_{121}	X_{131}	$\cdots$	X_{1n1}
	X_{112}	X_{122}	X_{132}		X_{1n2}
R_2	X_{211}	X_{221}	X_{231}	$\cdots$	X_{2n1}
	X_{212}	X_{222}	X_{232}		X_{2n2}
$\vdots$	$\vdots$	$\vdots$	$\vdots$		$\vdots$
R_n	X_{m11}	X_{m21}	X_{m31}	$\cdots$	X_{mn1}
	X_{m12}	X_{m22}	X_{m32}		X_{mn2}

Example: Let us assume that in the example shown in Table 5.11, we took three replicate measurements for each of the 10 paint type–drying time situations. Table 5.14 shows the resulting data.

TABLE 5.14

Drying Time, min	1	2	3	4	5	Σ
3	3.20	3.60	3.91	3.59	4.11	18.41
	3.28	3.52	3.82	3.68	4.06	18.36
	3.26	3.58	3.88	3.65	4.01	18.38
Σk	9.74	10.70	11.61	10.92	12.18	55.15
	4.08	4.69	5.14	4.80	4.99	23.70
5	4.10	4.68	5.06	4.71	4.89	23.44
	4.15	4.56	5.12	4.73	4.92	23.48
Σk	12.33	13.93	15.32	14.24	14.80	70.62
Σc	22.07	24.63	26.93	25.16	26.98	125.77

$$c = 5, \quad r = 2, \quad k = 3 \quad \therefore crk = 5 \cdot 2 \cdot 3 = 30 = N$$

$$\Sigma X = 125.77$$
$$\Sigma X^2 = 538.16$$
$$\Sigma c^2 = 22.07^2 + 24.63^2 + 26.93^2 + 25.16^2 + 26.98^2 = 3179.89$$
$$\Sigma r^2 = 55.15^2 + 70.62^2 = 8028.71$$
$$\Sigma k^2 = 9.74^2 + 10.70^2 + \cdots + 14.24^2 + 14.80^2 = 1614.34$$

The generalized format for two-factor ANOVA with replication is shown in Table 5.15. Note particularly the denominator of the first quantity in the SS_c and SS_r column. Each c which is squared is made up of six measurements: $(3.20 + 3.28 + 3.26 + 4.08 + 4.10 + 4.15 = 22.07)$ and $rk = 2 \cdot 3 = 6$. Similarly, each r which is squared is made up of 15 measurements, i.e., all 15 for row 3 and all 15 for row 5 and $ck = 5 \cdot 3 = 15$. Now, applying the example problem under discussion to Table 5.15, we get Table 5.16.

F tests:
$$F = 0.68/0.0025 = 272; \qquad F_{0.01;4,20} = 4.43$$
$$F = 7.98/0.0025 = 3192; \qquad F_{0.01;1,20} = 8.10$$
$$F = 0.04/0.0025 = 16; \qquad F_{0.01;4,20} = 4.43$$

From the above F tests it can be concluded that the effects of the P, D, and PD levels are all significant at the $\alpha = 0.01$ level. What we are really saying is that the sum of squares contribution is significant because our error term is so small. It can

TABLE 5-15

Source	SS	d.f.	MS
Column factor, c	$\sum \frac{\Sigma c^2}{rk} - \frac{(\Sigma X)^2}{N}$	$c - 1$	$SS_c/(c - 1)$
Row factor, r	$\sum \frac{\Sigma r^2}{ck} - \frac{(\Sigma X)^2}{N}$	$r - 1$	$SS_r/(r - 1)$
$r \cdot c$ interaction	By difference	$(c - 1)(r - 1)$	$SS_{rc}/[(c - 1)(r - 1)]$
Subtotal	$\sum \frac{\Sigma k^2}{k} - \frac{(\Sigma X)^2}{N}$	$rc - 1$	—
Error	$SS_{total} - SS_{subtotal}$	$rc(k - 1)$	$SS_E/[rc(k - 1)]$
Total	$\Sigma X^2 - (\Sigma X)^2/N$	$N - 1$	—

TABLE 5.16

Source	SS		d.f.	MS
Paint type, P	$\frac{3179.89}{6} - \frac{125.77^2}{30}$	= 2.71	4	2.71/4 = 0.68
Drying time, D	$\frac{8028.71}{15} - 527.27$	= 7.98	1	7.98/1 = 7.98
PD interaction	By difference	= 0.15	4	0.15/4 = 0.04
Subtotal	$\frac{1614.34}{3} - 527.27$	= 10.84	9	
Error	By difference	= 0.05	20	0.05/20 = 0.0025
Total	538.16 − 527.27	= 10.89	29	

be seen that as the error term increases, the F value decreases until at some point it becomes less than the tabular value, in which case we cannot make any claim of significance.

5.7 GENERALIZED FACTORIAL ANALYSIS

We are not limited in the number of levels, subgroups, or replications that we might consider in any experimental situation. Naturally, the more complicated the case becomes, the more tedious the calculations and the more samples are required. For a completely balanced experiment with, say, three factors, two levels, three subgroups, and four replications, we need $3 \times 2 \times 3 \times 4 = 72$ measurements. In the previous section we considered interactions, but only two-way interactions. When enough variables are considered, we can have three-way or higher interactions.

There are ways to handle experiments that are not completely balanced, but they are beyond the scope of this work. The procedure for analysis is very much akin to the two-factor case with subgroups except that the absence of subgroups obviates calculation of interim sum of squares (see Table 5.15).

Let us now formulate the three-factor ANOVA format with two- and three-way interactions and replications. By symmetry Table 5.17 will suffice to indicate to the student the manner in which it can be extended for an n-factor experiment.

TABLE 5.17

Source	SS	d.f.	MS
A	$\sum \frac{(\Sigma A)^2}{bck} - \frac{(\Sigma X)^2}{N}$	$a - 1$	$SS_A/(a - 1)$
B	$\sum \frac{(\Sigma B)^2}{ack} - \frac{(\Sigma X)^2}{N}$	$b - 1$	$SS_B/(b - 1)$
C	$\sum \frac{(\Sigma C)^2}{abk} - \frac{(\Sigma X)^2}{N}$	$c - 1$	$SS_C/(c - 1)$
AB	$\sum \frac{(\Sigma AB)^2}{ck} - \frac{(\Sigma X)^2}{N} - SS_A - SS_B$	$(a - 1)(b - 1)$	$SS_{AB}/[(a - 1)(b - 1)]$
AC	$\sum \frac{(\Sigma AC)^2}{bk} - \frac{(\Sigma X)^2}{N} - SS_A - SS_C$	$(a - 1)(c - 1)$	$SS_{AC}/[(a - 1)(c - 1)]$
BC	$\sum \frac{(\Sigma BC)^2}{ak} - \frac{(\Sigma X)^2}{N} - SS_B - SS_C$	$(b - 1)(c - 1)$	$SS_{BC}/[(b - 1)(c - 1)]$
ABC	$\sum \frac{(\Sigma ABC)^2}{k} - \frac{(\Sigma X)^2}{N} - SS_A - SS_B - SS_C - SS_{AB} - SS_{AC} - SS_{BC}$	$(a - 1)(b - 1)(c - 1)$	$SS_{ABC}/[(a - 1)(b - 1)(c - 1)]$
Error	By difference	By difference	—
Total	$\Sigma X^2 - \frac{(\Sigma X)^2}{N}$	$abck - 1$	—

Consider the following nomenclature:

Factor A at level a,
factor B at level b,
factor C at level c, and
k replicates, so that
total measurements $= N = abck$.

Note the symmetry in the first term in the SS column. The factor(s) in the numerator is mutually exclusive with the levels in the denominator. Similarly, the levels in the denominator are mutually exclusive with the term(s) in the degrees of freedom column. Also note that in computing the interaction sum of squares, the main effect and lower-order interaction sum of squares is subtracted. When we have no replication, we use the highest-order interaction sum of squares for the error term.

Example: The data given in Table 5.18 describe an experiment where two different alloys each were formed in two ways (cast or forged). These samples were, in turn, machined in two ways (milled or turned). The resulting specimens were stressed or nonstressed and three readings of the strength of the metals were made of each unique condition. Perform an analysis of variance determination on this data.

As in previous examples, the tedious part of the problem is not the actual testing but the computations necessary to complete the ANOVA table.

$$\Sigma X = 395.6$$
$$\Sigma X^2 = 3562.96$$

Alloy:

$$\Sigma A_{\mathrm{I}} = 16.5 + 26.9 + \cdots + 16.8 + 25.1 = 166.1$$
$$\Sigma A_{\mathrm{II}} = 21.4 + 36.5 + \cdots + 21.3 + 36.2 = 229.5$$

Forming:

$$\Sigma F_C = 16.5 + 26.9 + 15.8 + 24.0 + 21.4 + 36.5 + 21.2 + 34.5 = 196.8$$
$$\Sigma F_F = 14.4 + 26.6 + 16.8 + 25.1 + 21.1 + 37.3 + 21.3 + 36.2 = 198.8$$

TABLE 5.18

Alloy, A	Forming, F	Machining, M	Stress S	Replicates 1	2	3	Total
Alloy I	Cast	Milled	Stressed	5.3	5.4	5.8	16.5
			Nonstressed	9.3	8.6	9.0	26.9
		Turned	Stressed	4.9	5.3	5.6	15.8
			Nonstressed	8.2	8.0	7.8	24.0
	Forged	Milled	Stressed	5.0	4.6	4.8	14.4
			Nonstressed	9.0	8.7	8.9	26.6
		Turned	Stressed	5.5	5.4	5.9	16.8
			Nonstressed	8.5	8.0	8.6	25.1
Alloy II	Cast	Milled	Stressed	7.2	7.1	7.1	21.4
			Nonstressed	12.2	12.4	11.9	36.5
		Turned	Stressed	7.0	6.8	7.4	21.2
			Nonstressed	11.5	11.4	11.6	34.5
	Forged	Milled	Stressed	7.4	6.2	7.5	21.1
			Nonstressed	12.7	11.9	12.7	37.3
		Turned	Stressed	6.9	7.0	7.4	21.3
			Nonstressed	12.0	11.9	12.3	36.2

Machining:

$$\Sigma M_M = 16.5 + 26.9 + 14.4 + 26.6 + 21.4 + 36.5 + 21.1 + 37.3 = 200.7$$

$$\Sigma M_T = 15.8 + 24.0 + 16.8 + 25.1 + 21.2 + 34.5 + 21.3 + 36.2 = 194.9$$

Stressing:

$$\Sigma S_S = 16.5 + 15.8 + 14.4 + 16.8 + 21.4 + 21.2 + 21.1 + 21.3 = 148.5$$

$$\Sigma S_N = 26.9 + 24.0 + 26.6 + 25.1 + 36.5 + 34.5 + 37.3 + 36.2 = 247.1$$

$$(\Sigma X)^2/N = 395.6^2/48 = 156.499.36/48 = 3260.40$$

Now from Table 5.17 and referring to factor A at level a, factor F at level f, factor M at level m, factor S at level s, and k replicates,

$$\text{Total measurements} = N = afmsk = 2\cdot 2\cdot 2\cdot 2\cdot 3 = 48$$

Therefore for the main-factor totals,

$$SS_A = \frac{(\Sigma A)^2}{fmsk} - (\Sigma X)^2/N = \frac{166.1^2 + 229.5^2}{2\cdot 2\cdot 2\cdot 3} - 3260.40 = 83.74$$

$$SS_F = \frac{(\Sigma F)^2}{amsk} - (\Sigma X)^2/N = \frac{196.8^2 + 198.8^2}{2\cdot 2\cdot 2\cdot 3} - 3260.40 = 0.09$$

$$SS_M = \frac{(\Sigma M)^2}{afsk} - (\Sigma X)^2/N = \frac{200.7^2 + 194.9^2}{2\cdot 2\cdot 2\cdot 3} - 3260.40 = 0.70$$

$$SS_S = \frac{(\Sigma S)^2}{afmk} - (\Sigma X)^2/N = \frac{148.5^2 + 247.1^2}{2\cdot 2\cdot 2\cdot 3} - 3260.40 = 202.54$$

Now before calculating the first-order interactions, we need the following totals. How many two-way interactions will we have? Since we want two-way interactions for our four factors, we have $\binom{4}{2} = \frac{4!}{(4-2)!2!} = \frac{24}{4} = 6$.

AF: Alloy I—cast = 16.5 + 26.9 + 15.8 + 24.0 = 83.2
Alloy I—turned = 15.8 + 24.0 + 16.8 + 25.1 = 82.9
Alloy II—cast = 21.4 + 36.5 + 21.2 + 34.5 = 113.6
Alloy II—forged = 21.1 + 37.3 + 21.3 + 36.2 = 115.9

AM: Alloy I—milled = 16.5 + 26.9 + 14.4 + 26.6 = 84.4
Alloy I—turned = 15.8 + 24.0 + 16.8 + 25.1 = 81.7
Alloy II—milled = 21.4 + 36.5 + 21.1 + 37.3 = 116.3
Alloy II—turned = 21.2 + 34.5 + 21.3 + 36.2 = 113.2

AS: Alloy I—stressed = 16.5 + 15.8 + 14.4 + 16.8 = 63.5
Alloy I—nonstressed = 26.9 + 24.0 + 26.6 + 25.1 = 102.6
Alloy II—stressed = 21.4 + 21.2 + 21.1 + 21.3 = 85.0
Alloy II—nonstressed = 36.5 + 34.5 + 37.3 + 36.2 = 144.5

FM: Cast—milled = 16.5 + 26.9 + 21.4 + 36.5 = 101.3
Cast—turned = 15.8 + 24.0 + 21.2 + 34.5 = 95.5
Forged—milled = 14.4 + 26.6 + 21.1 + 37.3 = 99.4
Forged—turned = 16.8 + 25.1 + 21.3 + 36.2 = 99.4

FS: Cast—stressed = 16.5 + 15.8 + 21.4 + 21.2 = 74.9
Cast—nonstressed = 26.9 + 24.0 + 36.5 + 34.5 = 121.9
Forged—stressed = 14.4 + 16.8 + 21.1 + 21.3 = 73.6
Forged—nonstressed = 26.6 + 25.1 + 37.3 + 36.2 = 125.2

MS: Milled—stressed = 16.5 + 14.4 + 21.4 + 21.1 = 73.4
Milled—nonstressed = 26.9 + 26.6 + 36.5 + 37.3 = 127.3
Turned—stressed = 15.8 + 16.8 + 21.2 + 21.3 = 75.1
Turned—nonstressed = 24.0 + 25.1 + 34.5 + 36.2 = 119.8

It should be noted at this point that the totals under each interaction category equal ΣX. This is a check to assure the choice of the correct categories.

Now calculating the two-way interaction sum of squares,

$$SS_{AF} = \sum \frac{(\Sigma AF)^2}{msk} - \frac{(\Sigma X)^2}{N} - SS_A - SS_F$$

$$= \frac{83.2^2 + 82.9^2 + 113.6^2 + 115.9^2}{2 \cdot 2 \cdot 3} - 3260.40 - 83.74 - 0.09 = 0.14$$

$$SS_{AM} = \sum \frac{(\Sigma AM)^2}{fsk} - \frac{(\Sigma X)^2}{N} - SS_A - SS_M$$

$$= \frac{84.4^2 + 81.7^2 + 116.3^2 + 113.2^2}{2 \cdot 2 \cdot 3} - 3260.40 - 83.74 - 0.70 = 0.01$$

$$SS_{AS} = \sum \frac{(\Sigma AS)^2}{fmk} - \frac{(\Sigma X)^2}{N} - SS_A - SS_S$$

$$\text{SS}_{AS} = \frac{63.5^2 + 102.6^2 + 85.0^2 + 144.5^2}{2\cdot 2\cdot 3} - 3260.40 - 83.74$$
$$- 202.54 = 8.68$$

$$\text{SS}_{FM} = \sum \frac{(\Sigma\, FM)^2}{ask} - \frac{(\Sigma\, X)^2}{N} - \text{SS}_F - \text{SS}_M$$
$$= \frac{101.3^2 + 95.5^2 + 99.4^2 + 99.4^2}{2\cdot 2\cdot 3} - 3260.40 - 0.09$$
$$-0.70 = 0.69$$

$$\text{SS}_{FS} = \sum \frac{(\Sigma\, FS)^2}{amk} - \frac{(\Sigma\, X)^2}{N} - \text{SS}_F - \text{SS}_S$$
$$= \frac{74.9^2 + 121.9^2 + 73.6^2 + 125.2^2}{2\cdot 2\cdot 3} - 3260.40 - 0.09$$
$$-202.54 = 0.44$$

$$\text{SS}_{MS} = \sum \frac{(\Sigma\, MS)^2}{afk} - \frac{(\Sigma\, X)^2}{N} - \text{SS}_M - \text{SS}_S$$
$$= \frac{73.4^2 + 127.3^2 + 75.1^2 + 119.8^2}{2\cdot 2\cdot 3} - 3260.40 - 0.70$$
$$- 202.54 = 1.77$$

Note the necessity to carry the calculation at least to two decimal places in the sum of squares calculations. Next calculate the second-order (three-way) interaction terms. There are $\binom{4}{3} = 4$ such terms.

AFM: Alloy I—cast—milled = 16.5 + 26.9 = 43.4
Alloy I—cast—turned = 15.8 + 24.0 = 39.8
Alloy I—forged—milled = 14.4 + 26.6 = 41.0
Alloy I—forged—turned = 16.8 + 25.1 = 41.9
Alloy II—cast—milled = 21.4 + 36.5 = 57.9
Alloy II—cast—turned = 21.2 + 34.5 = 55.7
Alloy II—forged—milled = 21.1 + 37.3 = 58.4
Alloy II—forged—turned = 21.3 + 36.2 = 57.5

AMS: Alloy I—milled—stressed = 16.5 + 14.4 = 30.9
Alloy I—milled—nonstressed = 26.9 + 26.6 = 53.5
Alloy I—turned—stressed = 15.8 + 16.8 = 32.6
Alloy I—turned—nonstressed = 24.0 + 25.1 = 49.1
Alloy II—milled—stressed = 21.4 + 21.1 = 42.5
Alloy II—milled—nonstressed = 36.5 + 37.3 = 73.8

Alloy II—turned—stressed = 21.2 + 21.3 = 42.5
Alloy II—turned—nonstressed = 34.5 + 36.2 = 70.7

FMS: Cast—milled—stressed = 16.5 + 21.4 = 37.9
Cast—milled—nonstressed = 26.9 + 36.5 = 63.4
Cast—turned—stressed = 15.8 + 21.2 = 37.0
Cast—turned—nonstressed = 24.0 + 34.5 = 58.5
Forged—milled—stressed = 14.4 + 21.1 = 35.5
Forged—milled—nonstressed = 26.6 + 37.3 = 63.9
Forged—turned—stressed = 16.8 + 21.3 = 38.1
Forged—turned—nonstressed = 25.1 + 36.2 = 61.3

AFS: Alloy I—cast—stressed = 16.5 + 15.8 = 32.3
Alloy I—cast—nonstressed = 26.9 + 24.0 = 50.9
Alloy I—forged—stressed = 14.4 + 16.8 = 31.2
Alloy I—forged—nonstressed = 26.6 + 25.1 = 51.7
Alloy II—cast—stressed = 21.4 + 21.2 = 42.6
Alloy II—cast—nonstressed = 36.5 + 34.5 = 71.0
Alloy II—forged—stressed = 21.1 + 21.3 = 42.4
Alloy II—forged—nonstressed = 37.3 + 36.2 = 73.5

The symmetry of the system should begin to become apparent to the student. We are separating variances and assigning that variance to selected combinations of variables. The student is further cautioned that if a negative SS results from his calculation, he has made an error.

$$SS_{AFM} = \sum \frac{(\Sigma\, AFM)^2}{sk} - \frac{(\Sigma\, X)^2}{N} - SS_A - SS_F - SS_M - SS_{AF} - SS_{AM} - SS_{FM}$$

$$= \frac{43.4^2 + 39.8^2 + 41.0^2 + 41.9^2 + 57.9^2 + 55.7^2 + 58.4^2 + 57.5^2}{2 \cdot 3} - 3260.40 - 83.74 - 0.09 - 0.70 - 0.14 - 0.01 - 0.69 = 0.22$$

$$SS_{AMS} = \sum \frac{(\Sigma\, AMS)^2}{fk} - \frac{(\Sigma\, X)^2}{N} - SS_A - SS_M - SS_S - SS_{AM} - SS_{AS} - SS_{MS}$$

$$= \frac{30.9^2 + 53.5^2 + 32.6^2 + 49.1^2 + 42.5^2 + 73.8^2 + 42.5^2 + 70.7^2}{2 \cdot 3} - 3260.40 - 83.74 - 0.70 - 202.54 - 0.01 - 8.68 - 1.77$$

$$= 0.17$$

$$SS_{FMS} = \sum \frac{(\Sigma\, FMS)^2}{ak} - \frac{(\Sigma\, X)^2}{N} - SS_F - SS_M - SS_S - SS_{FM} - SS_{FS} - SS_{MS}$$

$$SS_{FMS} = \frac{37.9^2 + 63.4^2 + 37.0^2 + 58.5^2 + 35.5^2 + 63.9^2 + 38.1^2 + 61.3^2}{2 \cdot 3}$$
$$- 3260.40 - 0.09 - 0.70 - 202.54 - 0.69 - 0.44 - 1.77$$
$$= 0.03$$

$$SS_{AFS} = \sum \frac{(\Sigma\, AFS)^2}{mk} - \frac{(\Sigma\, X)^2}{N} - SS_A - SS_F - SS_S - SS_{AF}$$
$$- SS_{AS} - SS_{FS}$$
$$= \frac{32.3^2 + 50.9^2 + 31.2^2 + 51.7^2 + 42.6^2 + 71.0^2 + 42.4^2 + 73.5^2}{2 \cdot 3}$$
$$- 3260.40 - 83.74 - 0.09 - 202.54 - 0.14 - 8.68 - 0.44$$
$$= 0.01$$

Next we have the third-order (four-way) interaction term of which there is, of course, only one.

$$SS_{AFMS} = \sum \frac{(\Sigma\, AFMS)^2}{k} - \frac{(\Sigma\, X)^2}{N} - SS_A - SS_F - SS_S - SS_S$$
$$- SS_{AF} - SS_{AM} - SS_{AS} - SS_{FM} - SS_{FS} - SS_{AFM}$$
$$- SS_{AMS} - SS_{FMS} - SS_{AFS}$$
$$= \frac{16.5^2 + 26.9^2 + \cdots + 21.3^2 + 36.2^2}{3}$$
$$- 3260.40 - 83.74 - 0.09 - 0.70 - 202.54 - 0.14$$
$$-0.01 - 8.68 - 0.69 - 0.44 - 1.44 - 0.22 - 0.17$$
$$- 0.03 - 0.01 = 0.10$$

The error term can now be calculated by difference, so let us set up the ANOVA table as Table 5.19. The total SS is $\Sigma\, X^2 - [(\Sigma\, X)^2/N]$ or $3562.96 - 3260.40 = 302.56$. The fact that all degrees of freedom are 1 (except the error term) is the result of having only two levels of all main effects. Assuming this experiment to be of the model I type, the F tests are the ratios of the main effect and cross-product mean squares to the error mean square. Table 5.20 tabulates these F tests and lists the tabular values with which the calculated F values are compared. Again, each tabular F value would be different if the degrees of freedom associated with the numerator were other than 1 in each case.

From a comparison of the calculated F and the tabular F, it can be seen that the alloy and stress main effects, the alloy-stress cross product (as might be expected), and the machining-stress cross product are all statistically significant (with an α

TABLE 5.19

Source	SS	d.f.	MS
Alloy, A	83.74	$(a - 1) = 1$	83.74
Forming, F	0.09	$(f - 1) = 1$	0.09
Machining, M	0.70	$(m - 1) = 1$	0.70
Stress, S	202.54	$(s - 1) = 1$	202.54
AF	0.14	$(a - 1)(f - 1) = 1$	0.14
AM	0.01	$(a - 1)(m - 1) = 1$	0.01
AS	8.68	$(a - 1)(s - 1) = 1$	8.68
FM	0.69	$(f - 1)(m - 1) = 1$	0.69
FS	0.44	$(f - 1)(s - 1) = 1$	0.44
MS	1.77	$(m - 1)(s - 1) = 1$	1.77
AFM	0.22	$(a - 1)(f - 1)(m - 1) = 1$	0.22
AMS	0.17	$(a - 1)(m - 1)(s - 1) = 1$	0.17
FMS	0.03	$(f - 1)(m - 1)(s - 1) = 1$	0.03
AFS	0.01	$(a - 1)(f - 1)(s - 1) = 1$	0.01
$AFMS$	0.10	$(a - 1)(f - 1)(m - 1)(s - 1) = 1$	0.10
Subtotal	299.33		
Error (302.54 − 299.33)	3.23	Difference $= 32$	$3.23/32 = 0.10$
Total	302.56	$N - 1 = 47$	

TABLE 5.20

Source	MS	Calculated F		$F_{.01,1,32}$
A	83.74	83.74/0.10 =	837.4	7.50
F	0.09	0.09/0.10 =	0.9	7.50
M	0.70	0.70/0.10 =	7.0	7.50
S	202.54	202.54/0.10 =	2025.4	7.50
AF	0.14	0.14/0.10 =	1.4	7.50
AM	0.01	0.01/0.10 =	0.1	7.50
AS	8.68	8.68/0.10 =	86.8	7.50
FM	0.69	0.69/0.10 =	6.9	7.50
FS	0.44	0.44/0.10 =	4.4	7.50
MS	1.77	1.77/0.10 =	17.7	7.50
AFM	0.22	0.22/0.10 =	2.2	7.50
AMS	0.17	0.17/0.10 =	1.7	7.50
FMS	0.03	0.03/0.10 =	0.3	7.50
AFS	0.01	0.01/0.10 =	0.1	7.50
$AFMS$	0.10	0.10/0.10 =	1.0	7.50

of 0.01). Significance indicates that these main-effect and cross-product terms do, in fact (and again with a probability of error of 0.01), have an effect on the strength of the metal.

From the above it can be seen that the main purpose of this experiment (and, indeed, most engineering experiments) is that of considering the hypotheses about the relative effects of the treatment. Where the calculated F is less than the tabular F, the null hypothesis

$$H_o{:}\tau_i = 0$$

is accepted and we assume "no difference" among the various treatments.

6
Regression by Least Squares Method

"NEVER DESPAIR. BUT IF YOU DO, WORK ON IN DESPAIR." *Edmund Burke*

If the behavior of one variable can be described in terms of another, they are said to be "related," in a mathematical sense. At the very core of any engineering discipline is the appreciation of and manipulation of variables. A formula, i.e., mathematical model, is the method commonly used to express these relationships. Examples are

1. In inventory control

$$Q_m = \sqrt{\frac{2a_c s}{I}} \tag{6.1}$$

or

$$Q_m = 1.41 a_c^{1/2} s^{1/2} (1/I)^{1/2}$$

2. In machine design

$$\omega_{cr} = \sqrt{\frac{15.30\,Wg}{y}} \tag{6.2}$$

or

$$\omega_{cr} = 3.92\,W^{1/2} g^{1/2} (1/y)^{1/2}$$

3. In heat transfer

$$U = \frac{1}{(1/h_{12}) + (\Delta X_{23}/k_{23}) + (1/h_{34})} \tag{6.3}$$

or

$$U = h_{12} + k_{23}(1/\Delta X_{23}) + h_{34}$$

Now, what do all these models have in common? They define the relationship between the variables. In engineering, we use models to predict the future behavior of any engineering system based on one of two criteria. The first criterion is the pure physical, chemical, or mathematical base. Examples include Newton's, Einstein's, Boyle's, or Charles' laws. The second criterion, and by far the most common to practicing engineers, is that model resulting from empirical experimentation. Typical of these are most fluid flow, machine design, strength of material, production control systems design, and the like. These models result from the analysis of data gained either by experimentation or the collection of measured data from an actual system. The type of analysis to which the data are subjected, in most cases, is regression analysis. Regression analysis yields a model which can be used for predictive purposes whenever a situation with similar conditions and variables is encountered.

A word about prediction is now in order. If any single word could be found which would describe engineering, *prediction* would come closest to filling the bill. Implicit in energy design formula is the notion that a formula represents a physical system and that the formula has been proved by mathematical logic or experimentation. Therefore since it has been proved and has worked in the past, we can use it to *predict* future behavior. So, in using any design formula, we are, in effect, predicting future behavior. From this it logically follows that engineers should possess the methodology of creating mathematical models as well as the understanding of how they are evolved.

We have referred to the relationship between and among variables. In addition to seeking a manner of expressing the form of a functional relationship, it is helpful to know how strong the relationship is. In addition, we want to know how precisely the value of a variable can be predicted if we know the values of other variables. All these as well as other aims can be accomplished using the regression-correlation approach. The two ideas cannot, for our purposes, be separated. Regression has to do with the *nature* of the relationship between variables; correlation has to do with the *degree* of association between and among variables.

6.1 METHOD OF LEAST SQUARES

The method of least squares is but one of a number of processes in mathematics which are lumped under the title of finite differences. Finite differences are used to solve systems of differential equations complicated enough so that approximate numerical methods will suffice.

The least squares method is a problem of curve fitting. There are two interpretations to this problem: one where the equation of the curve passes exactly through each point of a set and a second where the equation of the curve comes as close as possible to each point in a given set. Certainly the latter case is superior to an irregular curve which is forced to pass through each and every point. Particularly, in experimental data various errors (in addition to measurement errors) are present which cause irregularities. These errors, partly discussed with ANOVA, will be elaborated on shortly. The determination of "as close as possible" is usually called the least squares method.

Assume that we want to fit a straight line, l, whose equation is

$$y = a + bx \tag{6.4}$$

to n points, the first point being $x_1 y_1$ and the last being $x_n y_n$. Generally, in any given averaging process such as hypothesizing a line among points, no point is exactly on the line. Therefore, no general point, $x_i y_i$, will satisfy (6.4). If we substitute x_i in (6.4), we get the ordinate of l, which differs from y_i by δ_i. Therefore

$$\delta_i = y_i - (a + bx_i) \tag{6.5}$$

Noting Figure 6.1, we describe the difference δ_i for each point and formulate the sum of squares for each point. From Figure 6.1 and from model (6.5), we get

$$D = \sum_{i=1}^{n} \delta_1^2 = (y_1 - a - bx_1)^2 + (y_2 - a - bx_2)^2 + \cdots + (y_n - a - bx_n)^2 \tag{6.6}$$

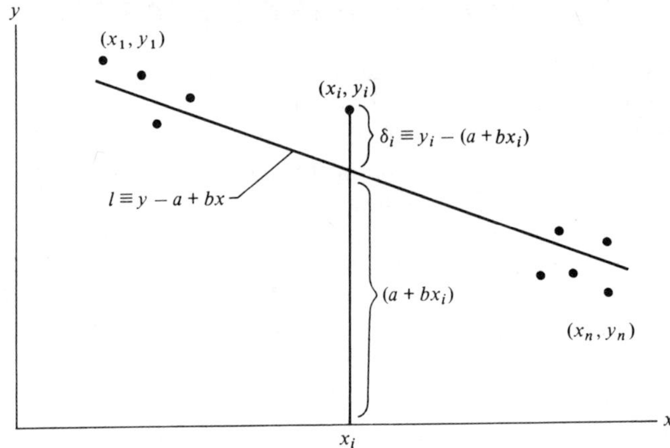

Fig. 6.1 Definition of terms used in development of least squares normal equations.

It can be seen that D is the measure of how well the curve (we will refer to straight lines as curves), l, fits the points. Obviously, D is zero only when each point lies on the curve l, and the larger D becomes, the farther the points are from l. In summary, the parameters a and b should be chosen so that the sum of squares of D are at a minimum.

This minimization is performed in the classic calculus manner, that of writing an alphanumeric expression containing the parameters a and b. Then equate to zero the partial derivatives. Performing these operations on model (6.6), we get

$$\frac{\partial D}{a} = 2(y_1 - a - bx_1)(-1) + 2(y_2 - a - bx_2)(-1) + \cdots + 2(y_n - a - bx_n)(-1) = 0 \tag{6.7}$$

$$\frac{\partial D}{b} = 2(y_1 - a - bx_1)(-x_1) + 2(y_2 - a - bx_2)(-x_2) + \cdots + 2(y_n - a - bx_n)(-x_n) = 0 \tag{6.8}$$

Collecting terms, we get

$$na + b\sum_{i=1}^{n} x_i = \sum_{i=1}^{n} y_i \tag{6.9}$$

$$a\sum_{i=1}^{n} x_i + b\sum_{i=1}^{n} x_i^2 = \sum_{i=1}^{n} x_i y_i \tag{6.10}$$

Some algebraic manipulation will yield Eqs. (6.9) and (6.10) in a form which is sometimes more convenient for computation. These are

$$b = \frac{n \Sigma xy - \Sigma x \Sigma y}{n \Sigma x^2 - (\Sigma x)^2} \tag{6.9a}$$

$$a = 1/n(\Sigma y - b \Sigma x) \tag{6.10a}$$

Note that the index and limit notations have been omitted for simplicity sake. This will be our habit in the future.

The normal equations can also be expressed in terms of sums of squares:

$$b = \frac{\mathrm{SS}xy}{\mathrm{SS}x^2} \tag{6.9b}$$

$$a = \bar{y} - b\bar{x} \tag{6.10b}$$

where

$$\mathrm{SS}xy = \Sigma xy - \frac{\Sigma x \Sigma y}{N} = \Sigma xy - \bar{x} \Sigma y$$

and

$$\mathrm{SS}x^2 = \Sigma x^2 - \frac{(\Sigma x)^2}{N} = \Sigma x^2 - \bar{x} \Sigma x$$

Equations (6.9) and (6.10) are known as the normal equations for *simple linear regression.*

Example: In problems of bearing heating, the relationship between the temperature rise of the oil film and the bearing wall is of interest. In a certain test the data in Table 6.1

TABLE 6.1

Sample	Temperature Rise of Oil Film, °F (y)	Temperature Rise of Bearing Wall, °F (x)
1	20	8
2	30	10
3	40	16
4	50	22
5	60	21
6	70	31
7	80	30
8	90	40
9	100	41
10	110	45

were collected. By consulting (6.9) and (6.10) it appears that the following information is necessary: n, Σx, Σy, Σx^2, and Σxy. For reasons which will become obvious later, we will add Σy^2 to the above list. Table 6.2 accumulates the necessary

TABLE 6.2

Sample	x	y	xy	x^2	y^2
1	8	20	160	64	400
2	10	30	300	100	900
3	16	40	640	256	1,600
4	22	50	1,100	484	2,500
5	21	60	1,260	441	3,600
6	31	70	2,170	961	4,900
7	30	80	2,400	900	6,400
8	40	90	3,600	1600	8,100
9	41	100	4,100	1681	10,000
10	45	110	4,950	2025	12,100
	264	650	20,680	8512	50,500

data, but first let us plot the information found in Table 6.1 as Figure 6.2. Note that $n = 10$. Now substituting the information from Table 6.2 in (6.9) and (6.10),

$$10a + 264b = 650 \tag{6.11}$$

$$264a + 8512b = 20{,}680 \tag{6.12}$$

Solving (6.11) and (6.12), we get $a = 4.81$ and $b = 2.28$, and since these data are of the form

$$y = a + bx$$

we have

$$y = 4.81 + 2.28x$$

for the equation of the curve which best represents the data shown in Table 6.1. Of course, negative values are accepted.

Two questions immediately become apparent. First, how do we know that the data are better represented by a straight line or a second- or higher-order equation? In answer to this we can only say *at this point* that we must make a judgment

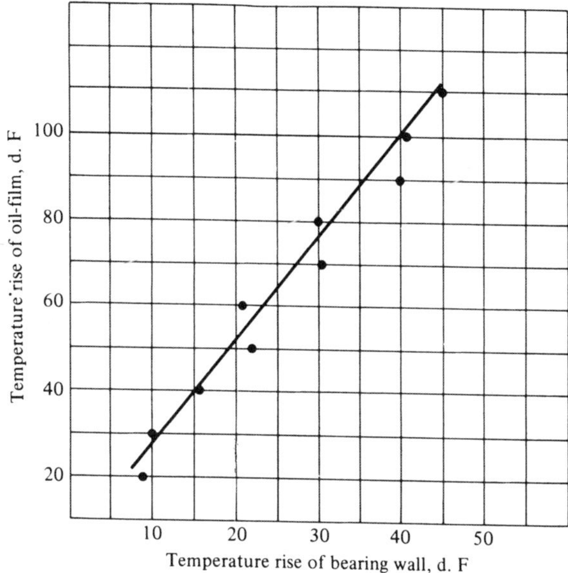

Fig. 6.2 Plot of bearing-wall temperature rise vs. oil-film temperature rise.

decision based on the plot of the data. Later on in this chapter we will consider a procedure for making this determination.

Second, how far can the points deviate from the hypothesized curve with the curve still considered to represent the data adequately. Here is where the *coefficient of correlation* enters the stage.

In the simple linear case the expression for the coefficient of correlation is

$$r = \frac{n\Sigma xy - \Sigma x \Sigma y}{\sqrt{[n\Sigma x^2 - (\Sigma x)^2][n\Sigma y^2 - (\Sigma y)^2]}} \tag{6.13}$$

where all terms are as computed in Table 6.2. The simple linear coefficient of correlation can also be expressed in terms of sum of squares:

$$r = \frac{\mathrm{SS}xy}{\sqrt{\mathrm{SS}x^2\,\mathrm{SS}y^2}} \tag{6.13a}$$

where terms are as previously defined and

$$SSy^2 = \Sigma y^2 - \frac{(\Sigma y)^2}{N} = \Sigma y^2 - \bar{y}\Sigma y$$

It should be emphasized that model (6.13) is applicable to the simple linear case only. The value of r can range from -1 to $+1$. A value of $+1$ denotes perfect correlation in the positive direction (see Figure 6.3a). Perfect correlation indicates

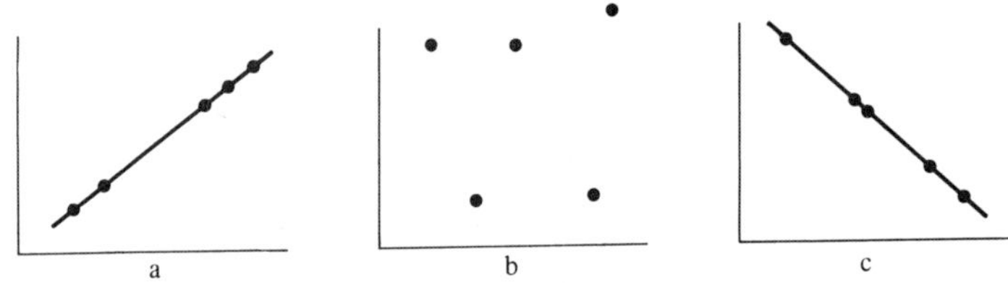

Fig. 6.3 Correlation coefficient relationships.

that all points are on the hypothesized curve. Figure 6.3b indicates what the data might look like to get a very low correlation figure, i.e., one approaching zero. A very low correlation indicates a lack of the hypothesized curve to represent the data. This does not rule out the possibility that some curve form might very well "fit" the data.

Figure 6.3c shows what the data would look like to get a perfectly negative correlation, i.e., an r value of -1.

Although we have a coefficient of correlation in the multilinear and curvilinear cases, its computation is somewhat more involved, though its meaning is the same in all three situations. Now, what about that meaning? If we square the coefficient of correlation, we get the product of correlation. The product of correlation, or r^2 value, indicates the percent of variation in the data that is represented by the hypothesized curve. For example, an r value of 0.9 will give an r^2 of 0.81. This indicates that 81% of the variation in the data is explained by the curve or model of the curve. The remaining 19% is usually due to inherent variations and does not necessarily cast a reflection on the accuracy or care with which the data are accumulated. Accordingly, it is not enough merely to present a model of the data. One must present the model and then qualify the presentation with the statement "with a coefficient (or product) of correlation of ________."

Example: Let us use the data from Table 6.1 to illustrate the simple linear correlation method.

Substitute the appropriate values from Table 6.2 in model (6.13). Now you can appreciate why we added the y^2 column to Table 6.2 even though we did not need it in models (6.9) and (6.10). Repeating (6.13),

$$r = \frac{n\Sigma xy - \Sigma x \Sigma y}{\sqrt{[n\Sigma x^2 - (\Sigma x)^2][n\Sigma y^2 - (\Sigma y)^2]}}$$

$$= \frac{10(20{,}680) - 264(650)}{\sqrt{(10.8512 - 264^2)(10.50500 - 650^2)}}$$

$$= \frac{35{,}200}{35{,}650} = 0.98$$

and

$$r^2 = 0.975$$

So, as we might expect from looking at Figure 6.2, there is strong evidence that the curve is a straight line. Since the r^2 value is 0.975, we can say that 97.5% of the variation in the data is explained by the line $y = 4.81 + 2.28x$.

If we use a sample of values to fit a least squares curve, then the function of the curve is an estimate of the true value and, in this case, y is $\hat{y}$, which denotes "estimated y." The true value of y falls within some range of values. These intervals are the confidence limits on the least squares line, as in the past these limits were a function of the t value and the variance. We can get limits on the slope and the mean value of y in addition to the estimated y.

Necessary to any consideration of these confidence limits are expressions for the variance of specific parameters, such as

$$\text{Variance of estimated } y = s^2(\hat{y}) = \frac{(1 - r^2)\,\text{SS}y^2}{N - 2} \qquad (6.13\text{b})$$

$$\text{Variance of slope} = s^2(b) = \frac{s^2(\hat{y})}{\text{SS}x^2} \qquad (6.13\text{c})$$

and

$$\text{Variance of mean of } y = s^2(\bar{y}) = \frac{s^2(\hat{y})}{N} \qquad (6.13\text{d})$$

The confidence limits of the, say, mean of the y's is

$$\bar{y} = \bar{y} \pm t_\alpha s(\bar{y}) \qquad (6.13e)$$

where α is the percent confidence range, usually 90 or 95%. It should be emphasized that the number of degrees of freedom attendant with t is $n - 1$, where n is the number of sample points used to calculate $s(\bar{y})$, and that if we are trying to get the limits for $\bar{y}$, we should use $s(\bar{y})$; if we are trying to get the limits for b, then we should use $s(b)$; etc.

Once we have computed $\bar{y}$ (or the limits on any other parameter) we can say that we are 95% sure, if we have used an α of 0.05, that the value for the average of the y's lies between $+ts(\bar{y})$ and $-ts(\bar{y})$.

Now, to compute the confidence limits for the example that we are now considering, we must generate SSx^2 and SSy^2. From Table 6.2 we get $\bar{x} = 26.4$ and $\bar{y} = 65.0$, so

$(x - \bar{x})$	$(x - \bar{x})^2$	$(y - \bar{y})$	$(y - \bar{y})^2$
−18.4	338.56	−45	2025
−16.4	268.96	−35	1225
−10.4	108.16	−25	625
−4.4	19.36	−15	225
−5.4	29.16	−5	25
4.6	21.16	5	25
3.6	12.96	15	225
13.6	184.96	25	625
14.6	213.16	35	1225
18.6	345.96	45	2025
	$SSx^2 = 1542.40$		$SSy^2 = 8250$

Therefore

$$s^2(\hat{y}) = \frac{(1 - r^2)\,SSy^2}{N\text{-}2} = \frac{(1 - 0.975)(8250)}{10\text{-}2} = 25.78$$

so

$$s^2(b) = \frac{s^2\hat{y}}{SSx^2} = \frac{25.78}{1542.40} = 0.017$$

and

$$s(b) = 0.13$$

Also,

$$s^2(\bar{y}) = \frac{s^2(\hat{y})}{N} = \frac{25.78}{10} = 2.578$$

We can now use this information to calculate the desired confidence limits:

$$t_{0.05,9} = 2.23$$

95% confidence limits for mean of y from (6.13e):

$$\begin{aligned} \bar{y} &= \bar{y} \pm ts(\bar{y}) \\ &= 65.0 \pm 2.23\sqrt{2.578} \\ &= 65.0 \pm 3.59 \end{aligned}$$

or 61.41 to 68.59; i.e., we are 95% confident that $\bar{y}$ will fall in this range. A similar process can yield confidence ranges on b.

6.2 LOGARITHMIC MODELS

It is well known that higher-order functions can be plotted on logarithmic or semilog plotting paper with a straight-line result. Since simple linear methods deal with fitting a curve to a set of data, a higher-order function can be accommodated using the logarithmic transform.

Example: Consider the data in Table 6.3, which are plotted

TABLE 6.3

x	y
1	1.5
2	1.7
3	2.1
4	2.2
5	2.7
6	3.0
7	3.5
8	4.0

in Figure 6.4. To use the simple linear expressions, it is necessary to transform the logarithmic scale (y) by substituting the natural logarithm for each value of y and thereafter accumulating the totals and products using ln y in place of y. If both x and y were logarithmic, then both scales would be transformed. A table of natural logarithms is included as Appendix 8. Table 6.4 accumulates the information generated from the

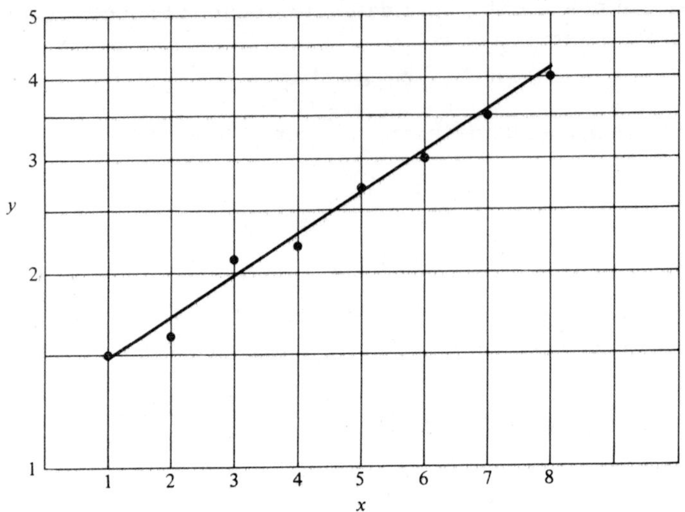

Fig. 6.4 Plot of data shown in Table 6.3.

data in Table 6.3 which is necessary to evolve a model and its attendant coefficient of correlation.

Model (6.9a):

$$b = \frac{n\Sigma xy - \Sigma x \Sigma y}{n\Sigma x^2 - (\Sigma x)^2}$$

$$= \frac{8(38.268) - 36(7.199)}{8(204) - (36)^2} = 0.140$$

TABLE 6.4

x	y	$\ln y$	$\ln y(x)$	x^2	$(\ln y)^2$
1	1.5	0.406	0.406	1	0.165
2	1.7	0.531	1.062	4	0.282
3	2.1	0.742	2.226	9	0.551
4	2.2	0.789	3.156	16	0.623
5	2.7	0.993	4.965	25	0.986
6	3.0	1.099	6.594	36	1.208
7	3.5	1.253	8.771	49	1.570
8	4.0	1.386	11.088	64	1.921
36		7.199	38.268	204	7.306

Model (6.10a):

$$a = 1/n(\Sigma y - b\Sigma x)$$
$$= \tfrac{1}{8}[7.199 - 0.140(36)] = 0.270$$

From $y = a + bx$ we get

$$\ln y = \ln a + bx \ln e$$

and since $\ln e = 1$,

$$\ln y = \ln a + bx$$

so

$$\ln y = 0.270 + 0.140x$$

and taking the antilog of all terms,

$$y = 1.3e^{0.14x} \qquad (\text{antilog of } 0.27 \simeq 1.3)$$

The coefficient of correlation from model (6.13) is (remembering to use $\ln y$ in place of y)

$$r = \frac{n\Sigma xy - \Sigma x \Sigma y}{\sqrt{[n\Sigma x^2 - (\Sigma x)^2][n\Sigma y^2 - (\Sigma y)^2]}}$$
$$= \frac{8(38.268) - 36(7.199)}{\sqrt{(8 \times 204 - 36^2)(8 \times 7.306 - 7.199^2)}} = 0.952$$

which might be expected from the apparent "good" fit of the data as seen in Figure 6.4.

6.3 MULTILINEAR MODELS

Up to this point we have been considering situations where we had only one dependent variable, y, and only one independent variable, x. We now consider the case where we have two or more independent variables, i.e., x's, but all variables are linear.

The least squares methodology referred to previously to develop the simple linear normal equations will pertain here with certain extensions. Again looking at the generalized problem, the method of least squares is a process of finding the best possible values for m unknowns, say, $x_1, x_2, \ldots, x_m$ related by n linear equations:

$$\begin{aligned} a_{11}x_1 + a_{12}x_2 + \cdots + a_{1m}x_m &= b_1 \\ a_{21}x_1 + a_{22}x_2 + \cdots + a_{2m}x_m &= b_2 \\ \vdots \qquad \vdots \qquad\qquad \vdots \quad &\quad \vdots \\ a_{m1}x_1 + a_{n2}x_2 + \cdots + a_{nm}x_m &= b_n \end{aligned} \tag{6.14}$$

restricted to the case where $n > m$. Enlarging on model (6.6), we attempt to find values for $x_1, x_2, \ldots, x_m$ which will result in

$$D = \sum_{i=1}^{n} \delta_i^2 = \sum_{i=1}^{n} (a_{i1}x_1 + a_{i2}x_2 + \cdots + a_{im}x_m - b_i)^2$$

being as small as possible.

Again following the notations used before, we want to minimize D and equate to zero each of its first partial derivatives, which are

$$\frac{\partial D}{\partial x_1}, \frac{\partial D}{\partial x_2}, \ldots, \frac{\partial D}{\partial x_m} \tag{6.15}$$

Now, developing $\partial D/\partial x_1$ we get

$$\frac{\partial D}{\partial x_1} = \sum_{i=1}^{n} 2(a_{i1}x_1 + a_{i2}x_2 + \cdots + a_{im}x_m - b_i)(a_{i1}) = 0$$

which, when terms are collected and arranged, becomes

$$x_1 \sum_{i=1}^{n} a_{i1}a_{i1} + x_2 \sum_{i=1}^{n} a_{i1}a_{i2} + \cdots + x_m \sum_{i=1}^{n} a_{i1}a_{im} = \sum_{i=1}^{n} a_{i1}b_i$$

Continuing this process we can generate a system of m linear equations whose solution can be obtained in the "usual manner." Now, as is our habit, we will restate the multilinear normal equations in the terms with which we have become familiar and eliminate the index and limiting restrictions on the summation signs. Prior to stating these equations, it is necessary for us to revise the notation that we have employed for the simple linear case. We do this for a number of reasons but mainly because problems of multivariable situations rapidly grow complex enough so that matrix notation becomes advantageous.

Therefore we will refer to the constant term (corresponding to the intercept or "a" term in the simple linear case) as b_0 and the coefficient attached to x_1 as b_1, etc. In the simple linear case we would have

$$y = b_0 + b_1 x$$

as the generalized case. In the multilinear situation we have

$$y = b_0 + b_1 x_1 + b_2 x_2 + \cdots + b_n x_n$$

as the general form, neglecting interaction terms. If interaction terms are considered, then we have

$$\begin{aligned} y = {} & b_0 + b_1 x_1 + b_2 x_2 + \cdots + b_n x_n + b_{n+1} x_1 x_2 \\ & + \cdots + b_{n+m} x_{n-1} x_n + \epsilon \end{aligned}$$

where there are n independent variables and m two-way interaction terms. ϵ is the error term which is made up of higher-order interaction terms and generally corresponds to the error term which was discussed in Chapter 5.

Now, to enumerate the normal equations for the multilinear regression case:

$$n(b_0) + \Sigma x_1(b_1) + \Sigma x_2(b_2) + \cdots = \Sigma y \qquad (6.16)$$

$$\Sigma x_1(b_0) + \Sigma x_1 x_1(b_1) + \Sigma x_1 x_2(b_2) + \cdots = \Sigma x_1 y \qquad (6.17)$$

$$\Sigma x_2(b_0) + \Sigma x_2 x_1(b_1) + \Sigma x_2 x_2(b_2) + \cdots = \Sigma x_2 y \qquad (6.18)$$

$$\vdots \qquad \vdots \qquad \vdots \qquad \vdots$$

$$\Sigma x_n(b_0) + \Sigma x_n x_1(b_1) + \Sigma x_n x_2(b_2) + \cdots + \Sigma x_n^2(b_n) = \Sigma x_n y \qquad (6.19)$$

As in the simple linear case, models (6.16) through (6.19) can be expressed in terms of sums of squares:

$$\mathrm{SS}x_1^2(b_1) + \mathrm{SS}x_1 x_2(b_2) + \mathrm{SS}x_1 x_3(b_3) + \cdots = \mathrm{SS}x_1 y \qquad (6.20)$$

$$\mathrm{SS}x_1 x_2(b_1) + \mathrm{SS}x_2^2(b_2) + \mathrm{SS}x_2 x_3(b_3) + \cdots = \mathrm{SS}x_2 y \qquad (6.21)$$

$$\mathrm{SS}x_1 x_3(b_1) + \mathrm{SS}x_2 x_3(b_2) + \mathrm{SS}x_3^2 + \cdots = \mathrm{SS}x_3 y \qquad (6.22)$$

$$\vdots$$

and

$$a\,(\text{or } b_0) = \bar{y} - b_1\bar{x}_1 - b_2\bar{x}_2 - b_3\bar{x}_3 - \cdots - b_n\bar{x}_n \qquad (6.23)$$

In models (6.20), (6.21), and (6.22)

$$SSx^2 = \Sigma(x - \bar{x})^2$$

and

$$SSxy = \Sigma(x - \bar{x})(y - \bar{y})$$

Noting models (6.16) through (6.19), it can easily be seen how, by symmetry, this system can grow. If we are to have n equations for n unknowns, each time we add an x, we add not only an equation but also a term to that portion of each equation to the left of the equal sign. In short, the system grows downward and outward. Further, the system shown in (6.16) through (6.18) is arranged so that this symmetry can more easily be appreciated, for instance, x_1^2 is written as x_1x_1, etc.

The coefficient of correlation of multilinear regression, denoted by R, is

$$R = \sqrt{\frac{b_1 SSx_1y + \cdots + b_n SSx_ny}{SS_y^2}} \tag{6.24}$$

where

$$SSx_1y = \Sigma(x_1 - \bar{x}_1)(y - \bar{y})$$

and

$$SSy^2 = \Sigma(y - \bar{y})^2$$

Example: Consider the data in Table 6.5 and compute the equation of the curve. The values to be calculated are seen in models (6.16) through (6.19). Substituting in (6.16) through (6.19),

$$\begin{aligned} 6b_0 + 21b_1 + 10.9b_2 &= 24.1 \\ 21b_0 + 91b_1 + 48.1b_2 &= 102.4 \\ 10.9b_0 + 48.1b_1 + 25.5b_2 &= 54 \end{aligned} \tag{6.25}$$

TABLE 6.5

n	y	x_1	x_2	x_1^2	x_1x_2	x_1y	y^2	x_2^2	x_2y
1	1.5	1.0	0.4	1	0.4	1.5	2.3	0.2	0.6
2	2.4	2.0	0.9	4	1.8	4.8	5.8	0.8	2.2
3	3.6	3.0	1.5	9	4.5	10.8	13.0	2.2	5.4
4	4.3	4.0	2.3	16	9.2	17.2	18.5	5.3	9.9
5	5.7	5.0	2.6	25	13.0	28.5	32.5	6.8	14.8
6	6.6	6.0	3.2	36	19.2	39.6	43.6	10.2	21.1
	24.1	21.0	10.9	91	48.1	102.4	115.7	25.5	54.0

There are a number of ways of solving the above system (including the computer), but we shall use a method known as Cramer's rule. The generalized form of Cramer's rule for solving a system of three equations with three unknowns is

$$\begin{aligned} ax_0 + bx_1 + cx_2 &= d \\ ex_0 + fx_1 + gx_2 &= h \\ ix_0 + jx_1 + kx_2 &= l \end{aligned}$$

where

$$x_0 = D_0/D, \quad x_1 = D_1/D, \quad \text{and} \quad x_2 = D_2/D \tag{6.26}$$

To get D first, write the matrix:

$$\begin{vmatrix} a & b & c \\ e & f & g \\ i & j & k \end{vmatrix} \tag{6.27}$$

Now rewrite the matrix repeating the first two columns:

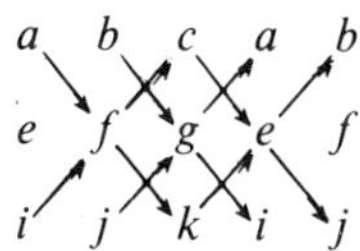

Following the diagonals,

$$D = (a \cdot f \cdot k) + (b \cdot q \cdot i) + (c \cdot e \cdot j) - (i \cdot f \cdot c) - (j \cdot g \cdot a) - (k \cdot e \cdot b)$$

To evaluate D_1 merely substitute the constant column, $\begin{matrix} d \\ h, \\ l \end{matrix}$ for the D_1 column in (6.27) and evaluate in the manner seen above. This process is continued until all unknowns in (6.26) equations are generated. Without completing this process, we will consider the case in our example. From (6.25),

$$D = \begin{matrix} 6 & 21 & 10.9 & 6 & 21 \\ 21 & 91 & 48.1 & 21 & 91 \\ 10 & 48.1 & 25.5 & 10 & 48.1 \end{matrix}$$

$$
\begin{aligned}
&= (6 \times 91 \times 25.5) + (21 \times 48.1 \times 10) \\
&\quad + (10.9 \times 21 \times 48.1) - (10 \times 91 \times 10.9) \\
&\quad - (48.1 \times 48.1 \times 6) - (25.5 \times 21 \times 21) \\
&= -13
\end{aligned}
$$

$$
D_0 = \begin{matrix} 24.1 & 21 & 10.9 & 24.1 & 21 \\ 102.4 & 91 & 48.1 & 102.4 & 91 \\ 54 & 48.1 & 25.5 & 54 & 48.1 \end{matrix}
$$

$$
\begin{aligned}
&= (24.1 \times 91 \times 25.5) + (21 \times 48.1 \times 54) \\
&\quad + (10.9 \times 102.4 \times 48.1) - (54 \times 91 \times 10.9) \\
&\quad - (48.1 \times 48.1 \times 24.1) - (25.5 \times 102.4 \times 21) \\
&= 0
\end{aligned}
$$

$$
D_1 = \begin{matrix} 6 & 24.1 & 10.9 & 6 & 24.1 \\ 21 & 102.4 & 48.1 & 21 & 102.4 \\ 10 & 54 & 25.5 & 10 & 54 \end{matrix}
$$

$$
\begin{aligned}
&= (6 \times 102.4 \times 25.5) + (24.1 \times 48.1 \times 10) \\
&\quad + (10.9 \times 21 \times 54) - (10 \times 102.4 \times 10.9) \\
&\quad - (54 \times 48.1 \times 6) - (25.5 \times 21 \times 24.1) \\
&= -30
\end{aligned}
$$

$$
D_2 = \begin{matrix} 6 & 21 & 24.1 & 6 & 21 \\ 21 & 91 & 102.4 & 21 & 91 \\ 10 & 48.1 & 54 & 10 & 48.1 \end{matrix}
$$

$$
\begin{aligned}
&= (6 \times 91 \times 54) + (21 \times 102.4 \times 10) \\
&\quad + (24.1 \times 21 \times 48.1) - (10 \times 91 \times 24.1) \\
&\quad - (48.1 \times 102.4 \times 6) - (54 \times 21 \times 21) \\
&= 33
\end{aligned}
$$

Therefore

$$
\begin{aligned}
x_0 &= D_0/D = 0/-13 = 0 \\
x_1 &= D_1/D = -30/-13 = 2.3076 \\
x_2 &= D_2/D = 33/-13 = -2.5384
\end{aligned}
$$

and the equation is $y = 2.3076x_1 - 2.5384x_2$. To calculate the coefficient of correlation (6.24), we need additional information from Table 6.5 (see Table 6.6).

TABLE 6.6

y	x_1	x_2	$x_1 - \bar{x}_1$	$y - \bar{y}$	$(x_1 - \bar{x}_1) \times (y - \bar{y})$	$(x_2 - \bar{x}_2)$	$(x_2 - \bar{x}_2) \times (y - \bar{y})$	$(y - y)^2$
1.5	1.0	0.4	−2.5	−2.52	6.30	−1.42	3.58	6.35
2.4	2.0	0.9	−1.5	−1.62	2.43	−0.92	1.49	2.62
3.6	3.0	1.5	−0.5	−0.42	0.21	−0.32	0.14	0.18
4.3	4.0	2.3	0.5	0.28	0.14	0.48	0.13	0.08
5.7	5.0	2.6	1.5	1.68	2.52	0.78	1.31	2.82
6.6	6.0	3.2	2.5	2.58	6.45	1.38	3.56	6.66
24.1	21.0	10.9			18.05		10.21	18.71

$$\bar{y} = 24.1/6 = 4.02$$
$$\bar{x}_1 = 21.0/6 = 3.50$$
$$\bar{x}_2 = 10.9/6 = 1.82$$

Now substituting in (6.24),

$$R = \sqrt{\frac{2.3076(18.05) - 2.5384(10.21)}{18.71}} = \sqrt{0.84} = 0.92$$

6.4 INTERACTIONS

We have considered the case where we can solve equations which result in the following model:

$$y = b_0 + b_1 x_1 + b_2 x_2 + \cdots + b_n x_n \tag{6.28}$$

How do we handle the case where we want to consider the effect of interactions? In model (6.28) we do not concern ourselves with cross products (interactions). Model (6.29) has been expanded to include them.

$$y = b_0 + b_1 x_1 + b_2 x_2 + \cdots + b_n x_n + b_{n+1} x_1 x_2 + \cdots + b_{n+m} x_{n-1} x_n \tag{6.29}$$

Model (6.29) for three independent variables would be

$$y = b_0 + b_1 x_1 + b_2 x_2 + b_3 x_3 + b_4 x_4 + b_5 x_5 + b_6 x_6 + \epsilon \tag{6.30}$$

We have used a very simple transform. When we get the equation we simply resubstitute the two-way interaction terms for x_4, x_5, and x_6.

6.5 CURVILINEAR REGRESSION

In the simple curvilinear case we have

$$y = b_0 + b_1 x_1 + b_2 x_1^2 + \epsilon \tag{6.31}$$

This case is handled similar to the interaction situation in that we would transform x_1^2 to x_2 and proceed as in the multilinear case.

If we hypothesize that a curve is a second degree, then any effects, as reflected in a sum of squares apportionment, due to third degree or higher will be included in the error term. We shall consider this problem in more detail later.

Example: Fit a parabola to the following data and compute the value of y for $x = 6$.

n	x	x'	y	$(x')^2$	$(x')^4$	$x'y$	x'^2y
1	0	−2.5	0.1	.	.	.	.
2	1	−1.5	0.9	.	.	.	
3	2	−0.5	3.9	.	.		
4	3	0.5	9.2	.			
5	4	1.5	15.9				
6	5	2.5	25.1				
	15	0	55.1	17.5	88.375	87.65	198.575

In the case of the parabola, the usual normal regression equations from (6.16) through (6.18) are

$$\begin{aligned} \Sigma y &= nb_0 + b_1 \Sigma x + b_2 \Sigma x^2 \\ \Sigma xy &= b_0 \Sigma x + b_1 \Sigma x^2 + b_2 \Sigma x^3 \\ \Sigma yx^2 &= b_0 \Sigma x^2 + b_1 \Sigma x^3 + b_2 \Sigma x^4 \end{aligned} \tag{6.32}$$

Fortunately, in most cases, the above equations can be simplified—if the values of the x's are placed at equal intervals (i.e., $x_1 = -4$, $x_2 = -3$, $x_3 = -2$, etc.), a particularly simple form for the normal equations is obtained.

Let us translate the y axis by means of the equation $x' = x - \bar{x}$. In effect this makes the mean of the x's $= 0$; that is, it distributes the values of x symmetrically about the origin.

Hence $x' = 0$, $x^3 = 0$, etc. (x^2's, not zero, because signs make all of them plus).

The normal equations then become

$$\Sigma y = nb_0 + b_2\Sigma(x')^2 \tag{6.33a}$$
$$\Sigma x'y = b_1\Sigma(x')^2 \tag{6.33b}$$
$$\Sigma(x')^2 y = b_0\Sigma(x')^2 + b_2\Sigma(x')4 \tag{6.33c}$$

Now, transforming x, $\bar{x} = \frac{15}{6} = 2.5$. Therefore $x' = x - \bar{x}$ so we get the x' column. Substituting in (6.33a),

$$55.1 = 6b_0 + 17.5b_2$$

Substituting in (6.33c),

$$198.575 = 17.5b_0 + 88.375b_2$$
$$b_0 = \frac{55.1 - 17.5b_2}{6}$$

And substituting this b_0 in (6.33c),

$$198.575 = 17.5\left(\frac{55.1 - 17.5b_2}{6}\right) + 88.375b_2$$

Therefore

$$b_2 = 1.0142$$

and substituting b_2 in (6.33a) we get

$$55.1 = 6b_0 + 17.5(1.0142)$$
$$b_0 = 6.2250$$

From (6.33b),

$$b_1 = 5.00857$$

Therefore

$$y = b_0 + b_1x' + b_2x'^2$$
$$= 6.2250 + 5.00857x' + 1.0142x'^2$$

Now for $x = 6$,

$$x' = 6 - 2.5 = 3.5$$
$$y = 6.2250 + 5.009(3.5) + 1.0142(3.5)^2$$
$$= 36.18$$

6.6 LINEAR OR CURVILINEAR?

Now that we are familiar with the different regression models, i.e., simple linear, multilinear (both with and without cross products), and curvilinear (again with and without cross products), we must develop a method for ascertaining which model best fits our data. Notice, we do not say "a method to determine whether our data are linear, quadratic, cubic, etc." Actually data seldom if ever exactly fit any one model; therefore we say "which best fits."

The method described above can best be accomplished using *orthogonal polynomials*. Orthogonal polynomials will be used only where the values of the independent variable (x's) are equally spaced. Actually there are methods of orthogonal polynomials for unequal intervals (see page 57, Reference 1), but that approach is outside the scope of this work. When we attempted to fit data to the simple linear normal equations or to second- or third-degree normal equations, we need, essentially, a new set of equations each time. Using orthogonal polynomials we can use the data from the previous results.

The most general independent case, as we have seen, is

$$y = b_0 + b_1 x_i + b_2 x_i^2 + b_3 x_i^3 + \cdots + b_{t-1} x_i^{t-1} + \epsilon_i \quad (6.34)$$

where $i = 1, 2, \ldots, t$ since interactions are not considered. If we have the values of the levels and the responses, we can estimate the coefficients b_0, b_1, b_2, etc., by the least squares method. From (6.34) for n levels, the degree of the polynomial cannot exceed $t - 1$.

Model (6.34) can be expressed in the following terms so as to obtain orthogonal polynomial nomenclature:

$$y = b_0 P_{0i} + b_1 P_{1i} + b_2 P_{2i} + \cdots + b_{t-1} P_{(t-1)i} + \epsilon \quad (6.35)$$

where $i = 1, 2, \ldots, t$ and $P_{0i} = 1$, and P_{mi} is a polynomial of the mth degree such that

$$P_{mi} = 0$$

The P_m's are the orthogonal polynomials. Reference 1 gives the method for calculating a small number of orthogonal polynomials. Table 6.7 is a partial table of orthogonal polynomials, where i is the number of data points, t is the number of treatments, i.e., groups of data, k is the order of hypothesized model, i.e., $k = 3$ indicates a test for a cubic equation.

TABLE 6.7 Orthogonal Polynomials

	$t = 2$	$t = 3$		$t = 4$			$t = 5$			
i	$k = 1$	$k = 1$	$k = 2$	$k = 1$	$k = 2$	$k = 3$	$k = 1$	$k = 2$	$k = 3$	$k = 4$
1	−1	−1	+1	−3	+1	−1	−2	+2	−1	+1
2	+1	0	−2	−1	−1	+3	−1	−1	+2	−4
3		+1	+1	+1	−1	−3	0	−2	0	+6
4				+3	+1	+1	+1	−1	−2	−4
5							+2	+2	+1	+1

The ANOVA method will be used for isolating linear, quadratic, or cubic components of treatment sums of squares. The tableau is shown in Table 6.8.

TABLE 6.8

Source / Treatment, T	d.f. / $(t - 1)$	SS / $(SS\,T)$	MS / MS
T_L	1	$SS\,T_L$	$SS\,T_L/1$
T_Q	1	$SS\,T_Q$	$SS\,T_Q/1$
T_C	1	$SS\,T_C$	$SS\,T_C/1$
T_{dev}	$(t - 4)$	$SS\,T_D$	$SS\,T_D/(t - 4)$
Error, E	$t(n - 1)$	$SS\,E$	$SS\,E/[t(n - 1)]$
Total	$tn - 1$	$x_i^2 - (x)^2/n$	

Example: Consider the data in Table 6.9, which show the effect of an additive on a certain type of jet fuel.

TABLE 6.9

	No Treatment	10 cu cm	20 cu cm	30 cu cm	40 cu cm	
	15	19	31	31	39	
	19	24	33	34	35	
	18	27	24	27	32	
	23	25	36	37	44	
	15	23	29	40	42	
Total	90	118	153	169	192	= 722
Mean	18.0	23.6	30.6	33.8	38.4	

$$\Sigma x^2 = 15^2 + 19^2 + \cdots + 44^2 + 42^2 = 22{,}532.00$$
$$(\Sigma x)^2/n = 442^2/25 = 20{,}851.36$$
$$\Sigma x^2 - (\Sigma x)^2/n = 22{,}532.00 - 20{,}851.36 = 1680.64$$

$$\text{SS}T = \frac{90^2 + 118^2 + 153^2 + 169^2 + 192^2}{5} - 20{,}851.36 = 1320.24$$

$$\text{SS}E = 22{,}532.00 - 20{,}851.36 - 1320.24 = 360.40$$

Now, consulting Table 6.7 for five treatments ($t = 5$) and five groups of data ($i = 5$), we get, for $k = 1$ (linearity), -2, -1, 0, $+1$, and $+2$. Therefore

$$\text{SS}T_L = \frac{[-2(90) + (-1)118 + 0(153) + 1(169) + 2(192)]^2}{5[(-2)^2 + (-1)^2 + 0^2 + (1)^2 + (2)^2]}$$
$$= 255^2/50 = 1300.50$$

To compute the quadratic contribution for five treatments and five groups, we consult the $k = 2$ column and get the following polynomials: $+2$, -1, -2, -1, and $+2$. Therefore

$$\text{SS}T_Q = \frac{[2(90) - 1(118) - 2(153) - 1(169) + 2(192)]^2}{5[2^2 + (-1)^2 + (-2)^2 + (-1)^2 + 2^2]}$$
$$= -29^2/70 = 12.00$$

To compute the cubic contribution for five treatments and five groups, we consult the $k = 3$ column and get the following polynomials: -1, $+2$, 0, -2, and $+1$. Therefore

$$\text{SS}T_C = \frac{[(-1)90 + 2(118) + 0(153) - 2(169) + 1(192)]^2}{5[-1^2 + 2^2 + 0^2 + (-2)^2 + 1^2]}$$
$$= 0^2/50 = 0$$

Now fitting the above data into the generalized ANOVA tableau (Table 6.8),

Source	d.f.	SS	MS	F Ratio
T	4	1320.24		
T_L	1	1300.50	1300.50	72.17
T_Q	1	12.00	12.00	0.67
T_C	1	0.00	0.00	0.00
T_{dev}	1	7.74	7.74	0.43
E	20	360.40	18.02	—

Consulting a table of F values for 1 and 20 d.f. and an α of 0.01, we get 8.10. Since the SST_{dev} term is less than this, we can be reasonably sure that it does not contribute to the equation. Comparing our F ratios with the tabular value of 8.10, we are strongly confident that our data are best represented by a linear relationship. The highest F ratio indicates the best fit and, therefore, the corresponding level (linear, quadratic, etc.) of the equation.

Example: This example is for the purpose of illustrating the computer solution of a multiple, curvilinear regression problem. There are eight independent variables (x's) and one dependent variable (y). The data are shown in Table 6.10. Although for illustrative purposes it is not necessary to define them, any or all of the independent variables could be cross-product or multinomial terms. For instance, x_4 could be x_2^2 or x_7 could be x_2x_3.

TABLE 6.10 Maintenance Costs Per Mile—Painting Stripes and Markings

Sample	Subsoil Factor, x_1	Surface-Condition ft, x_2	Surface Width, ft, x_3	Right-of-Way Width, ft, x_4	Traffic Volume, x_5	Maintenance Cost, \$ y
1	6.0	2.33	22	80	0.30	25
2	5.0	5.38	18	80	1.40	45
3	6.0	3.50	18	80	0.65	24
4	5.0	2.15	20	80	0.37	42
5	5.5	2.33	18	80	0.71	19
6	4.5	4.30	20	75	0.57	37
7	6.0	2.33	48	200	0.64	247
8	3.5	1.50	24	100	2.80	161
9	6.0	3.83	18	80	0.25	31
10	5.5	1.04	24	80	4.90	43
11	5.5	1.58	18	100	0.31	30
12	6.0	1.38	18	100	0.25	34
13	6.0	1.40	18	60	0.09	35
14	4.5	2.13	20	80	5.48	27
15	5.0	5.30	24	150	1.49	45
16	2.0	6.80	20	100	0.51	51
17	2.5	7.40	18	80	0.21	43

The information shown on pages 140 through 149 is but a portion of the printout of a typical multiple regression program.

The MS column is, of course, the sum of squares divided by the degrees of freedom. The F ratio is

$$F_{1,N-k-1} = \frac{[SS_{(x_1,\ldots,x_k)} - SS_{(x_1,\ldots,x_{k-1})}]/1}{SSE_{(x_1,\ldots,x_k)}/(N-k-1)}$$

where N is the number of data points and k is the number of independent variables. This F test is formulated because we really are seeking a difference between steps rather than ascertaining whether or not the SS removed by regression is different from the error SS. For instance, the F test for the first deletion (x_2) is

$$F_{1,8} = \frac{(47{,}125 - 47{,}095)/1}{7953/8} = 0.03$$

The remainder of the F tests are as shown on the appropriate printouts together with the tabular F value at the 0.05 level. We are really testing whether the contribution of the deleted variable is significant in terms of the error term. If it is not significant, then we can delete it. When we reach one that is significant, we stop and assume our model to include the remainder of the terms.

In the case under consideration, the model would be (from page 148)

$$y = 173.17 - 28.68x_1 + 0.0044x_4 - 1.77x_7$$

with an R^2 of 0.794.

To make the best use of the regression technique we must be able to assume no error in the measurement of the independent variable. For a given x, the y's should be normally and independently distributed. Our error terms, ϵ's, must also be normally and independently distributed about a mean of zero. Further, the variance should be the same for all x's.

Reference

1. K. C. Peng, "*The Design and Analysis of Scientific Experiments*, Addison-Wesley Publishing Company, Inc., Reading, Mass., 1967.

Machine run stepwise regression illustration using data shown in Table 6.10.

```
Number of observations =   17
Number of independent variables in analysis =   8
Number of dependent variables in analysis =   1
Number of independent variables in input data =   5
Number of dependent variables in input data =   1
Tolerance factor for matrix inversion =   0.0010000
Probability level for deleting variables =   0.999900

        Sums of x values
( 1)  0.815000E 02    ( 2)  0.546600E 02    ( 3)  0.982400E 04    ( 4)  0.172925E 06
( 5)  0.207600E 02    ( 6)  0.413250E 03    ( 7)  0.239117E 03    ( 8)  0.684088E 02

        Raw xx matrix
( 1, 1)  0.4132500E 03    ( 1, 2)  0.2396650E 03    ( 1, 3)  0.4608800E 05    ( 1, 4)  0.7954125E 06
( 1, 5)  0.9801999E 02    ( 1, 6)  0.2167625E 04    ( 1, 7)  0.9366962E 03    ( 1, 8)  0.3266441E 03
( 2, 2)  0.2391166E 03    ( 2, 3)  0.2825960E 05    ( 2, 4)  0.5534814E 06    ( 2, 5)  0.5241160E 02
( 2, 6)  0.1155747E 04    ( 2, 7)  0.1281131E 04    ( 2, 8)  0.1311070E 03
( 3, 3)  0.1034355E 08    ( 3, 4)  0.1599508E 09    ( 3, 5)  0.1100188E 05    ( 3, 6)  0.2264640E 06
```

```
( 3, 7)  0.1078507E 06     ( 3, 8)  0.3391986E 05
( 4, 4)  0.2978531E 10     ( 4, 5)  0.1920632E 06     ( 4, 6)  0.3853606E 07     ( 4, 7)  0.2347611E 07
( 4, 8)  0.5168401E 06
( 5, 5)  0.6840879E 02     ( 5, 6)  0.4781200E 03     ( 5, 7)  0.1886153E 03     ( 5, 8)  0.3115559E 03
( 6, 6)  0.1161955E 05     ( 6, 7)  0.4196393E 04     ( 6, 8)  0.1591141E 04
( 7, 7)  0.7679273E 04     ( 7, 8)  0.3335959E 03
( 8, 8)  0.1549353E 04

         xy cross products
Y( 1)
         x( 1)  0.407950E 04     x( 2)  0.275543E 04     x( 3)  0.899876E 06     x( 4)  0.159902E 08
         x( 5)  0.123030E 04     x( 6)  0.188007E 05     x( 7)  0.112198E 05     x( 8)  0.345807E 04

         Sums of y squared
  ( 1) 0.106945E 06

         Sums of y values
  ( 1)  0.939000E 03

         Weighted Degrees of Freedom = 17
```

Mean values of x

(1)	0.479412E 01	(2)	0.321529E 01	(3)	0.577882E 03	(4)	0.101721E 05
(5)	0.122118E 01	(6)	0.243088E 02	(7)	0.140657E 02	(8)	0.402405E 01

Sum (xx), corrected for means

(1, 1)	0.2252942E 02	(1, 2)	−0.2238147E 02	(1, 3)	−0.1009412E 04	(1, 4)	−0.3361029E 05
(1, 5)	−0.1505882E 01	(1, 6)	0.1864559E 03	(1, 7)	−0.2096568E 03	(1, 8)	−0.1315727E 01
(2, 2)	0.6336863E 02	(2, 3)	−0.3327449E 04	(2, 4)	−0.2523258E 04	(2, 5)	−0.1433790E 02
(2, 6)	−0.1729728E 03	(2, 7)	0.5123004E 03	(2, 8)	−0.8884737E 02		
(3, 3)	0.4666436E 07	(3, 4)	0.6002050E 08	(3, 5)	−0.9949578E 03	(3, 6)	−0.1234588E 05
(3, 7)	−0.3033058E 05	(3, 8)	−0.5612374E 04				
(4, 4)	0.1219527E 10	(4, 5)	−0.1910870E 05	(4, 6)	−0.3499970E 06	(4, 7)	−0.8469703E 05
(4, 8)	−0.1790182E 06						
(5, 5)	0.4305717E 02	(5, 6)	−0.2653117E 02	(5, 7)	−0.1033882E 03	(5, 8)	0.2280167E 03
(6, 6)	0.1573941E 04	(6, 7)	−0.1616250E 04	(6, 8)	−0.7179597E 02		
(7, 7)	0.4315935E 04	(7, 8)	−0.6286204E 03				
(8, 8)	0.1274072E 04						

Sum (xy), corrected for means

Y(1)

x(1)	−0.422176E 03	x(2)	−0.263731E 03	x(3)	0.357244E 06	x(4)	0.643866E 07
x(5)	0.836153E 02	x(6)	−0.402524E 04	x(7)	−0.198783E 04	x(8)	−0.323512E 03

Sum (yy), corrected for means

(1) 0.550791E 05

Mean values of y

(1) 0.552353E 02

Regression analysis on *y*(1)

ANOVA

Source	SS	d.f.	MS
Total	0.55079059E 05	16	3442
Regression	0.47125603E 05	8	5890
Error	0.79534560E 04	8	969

Variance = 0.99418201E 03
Standard deviation = 0.31530651E 02
Correlation coefficient = 0.92498609E 00
R^2 = 0.85559927E 00

	Coefficient	Standard Error	*T* Ratio	*F* Ratio	SS
(0)	0.1947576E 03	0.6033111E 03	0.3228145	0.1042092	
(1)	−0.4726097E 02	0.7008019E 02	−0.6743842	0.4547940	0.4521481E 03
(2)	−0.4565021E 01	0.2585468E 02	−0.1765646	0.0311751	0.3099368E 02
(3)	0.2315380E−01	0.2745049E−01	0.8434748	0.7114498	0.7073106E 03
(4)	0.2886812E−02	0.1692014E−02	1.7061393	2.9109114	0.2893976E 04
(5)	0.4242115E 02	0.2541785E 02	1.6689513	2.7853984	0.2769193E 04
(6)	0.2392247E 01	0.7960766E 01	0.3005047	0.0903030	0.8977767E 02
(7)	−0.1277584E 01	0.3550056E 01	−0.3598770	0.1295115	0.1287580E 03
(8)	−0.8200977E 01	0.4609823E 01	−1.7790220	3.1649191	0.3146506E 04

Required probability = 0.999900
Probability for *x*(2) = 0.159938

Delete x(2)

ANOVA

Source	SS	d.f.	MS	Calc. F	Tab. $F_{1,8;.05}$
Total	0.55079059E 05	16	3442		
Regression	0.47094601E 05	7	6728	0.03	5.32
Error	0.79844580E 04	9	887		

Variance = 0.88716199E 03
Standard deviation = 0.29785264E 02
Correlation coefficient = 0.92468178E 00
R^2 = 0.85503641E 00

	Coefficient	Standard Error	T Ratio	F Ratio	SS
(0)	0.2021609E 03	0.5172428E 03	0.3908433	0.1527585	
(1)	−0.5234099E 02	0.6038448E 02	−0.8670827	0.7518323	0.6669971E 03
(3)	0.2261514E–01	0.2577032E–01	0.8775654	0.7701209	0.6832220E 03
(4)	0.2887666E–02	0.1598346E–02	1.8066591	3.2640172	0.2895712E 04
(5)	0.4169779E 02	0.2369690E 02	1.7596308	3.0963007	0.2746920E 04
(6)	0.2860302E 01	0.7090916E 01	0.4033755	0.1627118	0.1443517E 03
(7)	−0.1887932E 01	0.7635998E 00	−2.4724109	6.1128154	0.5423057E 04
(8)	−0.8035445E 01	0.4263639E 01	−1.8846449	3.5518862	0.3151098E 04

Required probability = 0.999900
Probability for x(6) = 0.304331

Delete $x($ 6)

ANOVA

Source	SS	d.f.	MS	Calc. F	Tab. $F_{1,9;05}$
Total	0.55079059E 05	16	3442		
Regression	0.46950311E 05	6	7825	0.16	5.12
Error	0.81287476E 04	10	813		

Variance = 0.81287476E 03
Standard deviation = 0.28510958E 02
Correlation coefficient = 0.92326418E 00
R^2 = 0.85241673E 00

	Coefficient	Standard Error	T Ratio	F Ratio	SS
(0)	0.1589967E 03	0.1271150E 03	1.2508104	1.5645267	
(1)	−0.2830589E 02	0.9254709E 01	−3.0585393	9.3546625	0.7604169E 04
(3)	0.2183104E–01	0.2459751E–01	0.8875302	0.7877099	0.6403095E 03
(4)	0.2795023E–02	0.1514086E–02	1.8460131	3.4077642	0.2770085E 04
(5)	0.3808570E 02	0.2100119E 02	1.8135022	3.2887902	0.2673374E 04
(7)	−0.1807700E 01	0.7056980E 00	−2.5615779	6.5616815	0.5333825E 04
(8)	−0.7502246E 01	0.3880136E 01	−1.9335009	3.7384257	0.3038872E 04

Required probability = 0.999900
Probability for $x($ 3) = 0.600643

Delete $x(\ 3)$

ANOVA

Source	SS	d.f.	MS	Calc. F	Tab. $F_{1,10;.05}$
Total	0.55079059E 05	16	3442		
Regression	0.46310000E 05	5	9262	0.79	4.96
Error	0.87690586E 04	11	797		

Variance = 0.79718714E 03
Standard deviation = 0.28234502E 02
Correlation coefficient = 0.91694679E 00
R^2 = 0.84079142E 00

	Coefficient	Standard Error	T Ratio	F Ratio	SS
(0)	0.1748235E 03	0.1148371E 03	1.5223609	2.3175827	
(1)	−0.2995899E 02	0.8977439E 01	−3.3371424	11.1365194	0.8877890E 04
(4)	0.3867722E–02	0.9031124E–03	4.2826590	18.3411677	0.1462134E 05
(5)	0.3191697E 02	0.1962554E 02	1.6262979	2.6448447	0.2108436E 04
(7)	−0.2014510E 01	0.6596586E 00	−3.0538680	9.3261095	0.7434665E 04
(8)	−0.6447437E 01	0.3657825E 01	−1.7626424	3.1069080	0.2476787E 04

Required probability = 0.999900
Probability for $x(\ 5)$ = 0.870795

Delete *x*(5)

ANOVA

Source	SS	d.f.	MS	Calc. F	Tab. $F_{1,11;.05}$
Total	0.55079059E 05	16	3,442		
Regression	0.44201556E 05	4	11,051	2.65	4.84
Error	0.10877492E 05	12	906		

Variance = 0.90645768E 03
Standard deviation = 0.30107436E 02
Correlation coefficient = 0.89582995E 00
R^2 = 0.80251129E 00

	Coefficient	Standard Error	T Ratio	F Ratio	SS
(0)	0.1915921E 03	0.1184397E 03	1.6176344	2.6167409	
(1)	−0.3095652E 02	0.9550587E 01	−3.2413210	10.5061616	0.9523391E 04
(4)	0.4189962E−02	0.9395562E−03	4.4595112	19.8872392	0.1802694E 05
(7)	−0.1980345E 01	0.7030602E 00	−2.8167499	7.9340802	0.7191908E 04
(8)	−0.6742538E 00	0.9405189E 00	−0.7168955	0.5139392	0.4658641E 03

Required probability = 0.999900
Probability for *x*(8) = 0.506927

Delete $x(\ 8)$

ANOVA

Source	SS	d.f.	MS	Calc. F	Tab. $F_{1,12;.05}$
Total	0.55079059E 05	16	3,442		
Regression	0.43735705E 05	3	14,579	0.51	4.75
Error	0.11343352E 05	13	873		

Variance = 0.87256561E 03
Standard deviation = 0.29539222E 02
Correlation coefficient = 0.89109666E 00
R^2 = 0.79405325E 00

	Coefficient	Standard Error	T Ratio	F Ratio	SS
(0)	0.1731688E 03	0.9468280E 02	1.8289363	3.3450078	
(1)	−0.2867730E 02	0.8835908E 01	−3.2455405	10.5335333	0.9191199E 04
(4)	0.4366503E–02	0.8895975E–03	4.9084033	24.0924230	0.2102222E 05
(7)	−0.1767956E 01	0.6255580E 00	−2.8262064	7.9874425	0.6969568E 04

Required probability = 0.999900
Probability for $x(\ 7)$ = 0.986183

Delete *x*(7)

ANOVA

Source	SS	d.f.	MS	Calc. *F*.	Tab. $F_{1,13;.05}$
Total	0.55079059E 05	16	3,442		
Regression	0.36766150E 05	2	18,383	7.98	4.67
Error	0.18312908E 05	14	1,308		

Variance = 0.13080648E 04
Standard deviation = 0.36167180E 02
Correlation coefficient = 0.81701649E 00
R^2 = 0.66751595E 00

	Coefficient	Standard Error	*T* Ratio	*F* Ratio	SS
(0)	0.5901576E 02	0.4830618E 02	1.2217020	1.4925558	
(1)	−0.1132835E 02	0.7781387E 01	−1.4558264	2.1194305	0.2772353E 04
(4)	0.4967428E–02	0.1057635E–02	4.6967335	22.0593050	0.2885500E 05

Required probability = 0.999900
Probability for *x*(1) = 0.835426

Selected Readings

These references have been selected with the practicing engineer in mind. They are classified into four categories. The first category, "For Those Who Do Not Remember Calculus," includes works which require only facility in algebra. The second grouping, "For Those Who Remember Calculus," lists works for engineers who are recent graduates or whose work experience requires them to remain mathematically current. The third category, "For The Mathematically Sophisticated," includes references using matrices, determinants, higher-order differential equations, and transforms such as the LaPlace, Fourier, etc. The last category includes handbooks which would be useful in applied statistical problems.

FOR THOSE WHO DO NOT REMEMBER CALCULUS

Bartee, E. M., *Engineering Experimental Design Fundamentals*, Prentice-Hall, Inc., Englewood Cliffs, N.J., 1968.

Bowker, A. H., and C. J. Leiberman, *Handbook of Industrial Statistics*, Prentice-Hall, Inc., Englewood Cliffs, N.J., 1955. (A must in any plant library.)

Cowden, D. J., *Statistical Methods in Quality Control*, Prentice-Hall, Inc., Englewood Cliffs, N.J., 1957.

Dixon, W. J., and F. J. Massey, Jr., *Introduction to Statistical Analysis*, 2nd ed., McGraw-Hill, Inc., New York, 1957.

Duncan, A. J., *Quality Control and Industrial Statistics*, 3rd ed., Richard D. Irwin, Inc., Homewood, Ill., 1965. (A good handbook of statistical and quality-control methods.)

Ezekiel, M., and K. A. Fox, *Methods of Correlation and Regression Analysis*, 3rd ed., John Wiley & Sons, Inc., New York, 1959. (A very good, readable reference in the field.

Freund, J. E., *et al.*, *Manual of Experimental Statistics*, Prentice-Hall, Inc., Englewood Cliffs, N.J., 1960.

Hall, P. G., *Elementary Statistics*, John Wiley & Sons, Inc., New York, 1960.

Li, C. C., *Introduction to Experimental Statistics*, McGraw-Hill, Inc., New York, 1964.

Spregel, M. R., *Theory and Problems of Statistics*, Schaum Publishing Co., New York, 1961. (This is valuable for the self-teaching beginner and can be found in any college bookstore.)

Volk, William, *Applied Statistics for Engineers*, 2nd ed., McGraw-Hill, Inc., New York, 1969.

FOR THOSE WHO REMEMBER CALCULUS

Bowker, A. H., and C. J. Lieberman, *Engineering Statistics*, Prentice-Hall, Inc., Englewood Cliffs, N.J., 1959.

Guttman, I., and S. S. Wilks, *Introductory Engineering Statistics*, John Wiley & Sons, Inc., New York, 1965.

Hahn, G. J., and S. S. Shapiro, *Statistical Models in Engineering*, John Wiley & Sons, Inc., New York, 1967.

Johnson, N. L., and F. C. Leone, *Statistics and Experimental Design in Engineering and Physical Sciences*, Vols. I and II, John Wiley & Sons, Inc., New York, 1964.

FOR THE MATHEMATICALLY SOPHISTICATED

Brownlee, K. A., *Statistical Theory and Methodology in Science and Engineering*, 2nd ed., John Wiley & Sons, Inc., New York, 1965.

Draper, N. R., and H. Smith, *Applied Regression Analysis*, John Wiley & Sons, Inc., New York, 1966.

Williams, E. J., *Regression Analysis*, John Wiley & Sons, Inc., New York, 1959.

HANDBOOKS

Arkin, H., and R. R. Colton, *Tables for Statisticians*, Barnes & Noble, Inc., New York, 1950. (An excellent paperback reference.)

Beyer, W. H., ed., *Handbook of Tables for Probability and Statistics*, The Chemical Rubber Co., Cleveland, Ohio, 1966.

Glossary

Analysis of variance a computational process for separating a total variance into its components parts for the purposes of statistical inference.

Arithmetic mean the addition of a set of numbers divided by the number of items in that set, sometimes called the expected value.

Bayes' theorem a special case of conditional probability used to establish the probability that an event, previously occurred, might have happened in a particular manner.

Bias difference between the mean value of a statistic from a sample and the parameter from the large population that it is used to estimate.

Chauvenet's criterion a special case of the t distribution used to determine whether or not to reject an unwanted observation.

Chi-square tests of goodness of fit a test to determine the difference between a theoretical distribution and an observed one.

Class boundaries the most extreme true values included in any grouping of data.

Coefficient of correlation an indication of how well a group of data fits a theoretical mathematical model.

Coefficient of variation a measure of variability expressed in percent terms.

Combinations any of the different sets into which items may be grouped without regard to order within the group.

Conditional probability the probability that event B will occur first given that event A has occurred.

Confidence interval an interval that has a specified probability of including the large population value.

Confidence limits the end points of any confidence interval.

Confounding the mixing or ignoring of two or more factors so that their effects cannot be determined.

Cross product see *interaction*.

Degrees of freedom the number of independent observations in a sample minus the number of parameters which are estimated from sample observations.

Empirical probability the relative frequency of a probability distribution as contrasted with the classic definition which equates the theoretical frequency distribution.

Error mean square the residual mean square when the residual variation is associated with the experimental error.

Expected value the mean value of a random variable.

Factorial experiment an experiment where the levels of any factor are combined with all levels of the other factors.

Homogeneity of variances the characteristic which is denoted by the fact that variances of discrete positions on the random variable scale are from the same population.

Independence (probabilistic) the situation where the probability of event A in no way effects the outcome of event B.

Interaction the degree of mathematical relationship between two or more variables.

Marginal probability the cumulative probability of any row or column as compared with the probability in any or a group of cells.

Median that point in an array of data above and below which 50% of the data points lie.

Method of least squares the technique of estimating a function which minimizes the sums of squares of deviations from the estimate.

Mode the most frequently appearing value in any distribution.

Model I an analysis of variance or regression problem in which the sample data essentially represent the entire population.

Model II an analysis of variance or regression problem in which the data represent but a small sample of the large population.

Mutually exclusive the occurrence of an event excludes the occurrence of the remaining one or ones.

Normal equations the group of n equations for n unknowns used to ascertain the parameters in least squares regression problems.

Null hypothesis the hypothesis of no difference between two or more distributions.

One-tail test the test of the hypothesis that there is only one critical value; this can be either the upper or the lower limit.

Orthogonal polynomials the polynomials used to examine the numerical relationship between the response variable and the fixed levels of the factor.

Parameter a constant or coefficient that describes some characteristic of a distribution.

Permutations an arrangement of items as contrasted to combination which are the grouping of items.

PERT a critical-path-scheduling technique where the probabilistic nature of the time distribution is considered.

Power of a test the probability of rejecting a hypothesis when it is false $(1 - \beta)$.

Probability density function the function of a distribution where the random variable is continuous.

Probability distribution the distribution of the relative frequencies of a random variable; the random variable may be either discrete or continuous.

Probability mass function the function of a distribution where the random variable is discrete.

Process capability the actual capability of a process to meet specifications.

Randomness the condition which exists when there is no mathematical functional relationship among the items of data.

Random variable a statistic obtained from a random sample.

Regression analysis a term given to the technique of obtaining a mathematical function by the least squares method.

Replication obtaining repeated data points with no change of variables to determine the variance of the measuring technique.

Residual sum of squares the sum of squares remaining after those which are attributed to specific variables are apportioned.

Significance any result which deviates from some hypothetical value by more than can reasonably be attributed to chance errors.

Skewness that amount by which any distribution deviates from the symmetrical.

Standard deviation a measure of variability obtained by taking the root mean square of the deviations divided by the degrees of freedom.

Standard error the standard deviation of a sampling distribution.

Statistic a sample observation used to estimate a parameter.

Sum of squares the sum of the individual data points minus the mean, squared.

Two-tail test a test of hypothesis that has two critical values, one at the upper and one at the lower acceptance limit.

Variance the square of the standard deviation.

Appendix 1

FOUR-PLACE COMMON LOGARITHMS

N	0	1	2	3	4	5	6	7	8	9
10	0000	0043	0086	0128	0170	0212	0253	0294	0334	0374
11	0414	0453	0492	0531	0569	0607	0645	0682	0719	0755
12	0792	0828	0864	0899	0934	0969	1004	1038	1072	1106
13	1139	1173	1206	1239	1271	1303	1335	1367	1399	1430
14	1461	1492	1523	1553	1584	1614	1644	1673	1703	1732
15	1761	1790	1818	1847	1875	1903	1931	1959	1987	2014
16	2041	2068	2095	2122	2148	2175	2201	2227	2253	2279
17	2304	2330	2355	2380	2405	2430	2455	2480	2504	2529
18	2553	2577	2601	2625	2648	2672	2695	2718	2742	2765
19	2788	2810	2833	2856	2878	2900	2923	2945	2967	2989
20	3010	3032	3054	3075	3096	3118	3139	3160	3181	3201
21	3222	3243	3263	3284	3304	3324	3345	3365	3385	3404
22	3424	3444	3464	3483	3502	3522	3541	3560	3579	3598
23	3617	3636	3655	3674	3692	3711	3729	3747	3766	3784
24	3802	3820	3838	3856	3874	3892	3909	3927	3945	3962
25	3979	3997	4014	4031	4048	4065	4082	4099	4116	4133
26	4150	4166	4183	4200	4216	4232	4249	4265	4281	4298
27	4314	4330	4346	4362	4378	4393	4409	4425	4440	4456
28	4472	4487	4502	4518	4533	4548	4564	4579	4594	4609
29	4624	4639	4654	4669	4683	4698	4712	4728	4742	4757
30	4771	4786	4800	4814	4829	4843	4857	4871	4886	4900
31	4914	4928	4942	4955	4969	4983	4997	5011	5024	5038
32	5051	5065	5079	5092	5105	5119	5132	5145	5159	5172
33	5185	5198	5211	5224	5237	5250	5263	5276	5289	5302
34	5315	5328	5340	5353	5366	5378	5391	5403	5416	5428
35	5441	5453	5465	5478	5490	5502	5514	5527	5539	5551
36	5563	5575	5587	5599	5611	5623	5635	5647	5658	5670

N	0	1	2	3	4	5	6	7	8	9
37	5682	5694	5705	5717	5729	5740	5752	5763	5775	5786
38	5798	5809	5821	5832	5843	5855	5866	5877	5888	5899
39	5911	5922	5933	5944	5955	5966	5977	5988	5999	6010
40	6021	6031	6042	6053	6064	6075	6085	6096	6107	6117
41	6128	6138	6149	6160	6170	6180	6191	6201	6212	6222
42	6232	6243	6253	6263	6274	6284	6294	6304	6314	6325
43	6336	6345	6355	6365	6375	6385	6395	6405	6415	6425
44	6435	6444	6454	6464	6474	6484	6493	6503	6513	6522
45	6532	6542	6551	6561	6571	6580	6590	6599	6609	6618
46	6628	6637	6646	6656	6665	6675	6684	6693	6702	6712
47	6721	6730	6739	6749	6758	6767	6776	6785	6794	6803
48	6812	6821	6830	6839	6848	6857	6866	6875	6884	6893
49	6902	6911	6920	6928	6937	6946	6955	6964	6972	6981
50	6990	6998	7007	7016	7024	7033	7042	7050	7059	7067
51	7076	7084	7093	7101	7110	7118	7126	7135	7143	7152
52	7160	7168	7177	7185	7193	7202	7210	7218	7226	7235
53	7243	7251	7259	7267	7275	7284	7292	7300	7308	7316
54	7324	7332	7340	7348	7356	7364	7372	7380	7388	7396
55	7404	7412	7419	7427	7435	7443	7451	7459	7466	7474
56	7482	7490	7497	7505	7513	7520	7528	7536	7543	7551
57	7559	7566	7574	7582	7589	7597	7604	7612	7619	7627
58	7634	7642	7649	7657	7664	7672	7679	7686	7694	7701
59	7709	7716	7723	7731	7738	7745	7752	7760	7767	7774
60	7782	7789	7796	7803	7810	7818	7825	7832	7839	7846
61	7853	7860	7868	7875	7882	7889	7896	7903	7910	7917
62	7924	7931	7938	7945	7952	7959	7966	7973	7980	7987
63	7993	8000	8007	8014	8021	8028	8035	8041	8048	8055
64	8062	8069	8075	8082	8089	8096	8102	8109	8116	8122
65	8129	8136	8142	8149	8156	8162	8169	8176	8182	8189
66	8195	8202	8209	8215	8222	8228	8235	8241	8248	8254
67	8261	8267	8274	8280	8287	8293	8299	8306	8312	8319
68	8325	8331	8338	8344	8351	8357	8363	8370	8376	8382
69	8388	8395	8401	8407	8414	8420	8426	8432	8439	8445
70	8451	8457	8463	8470	8476	8482	8488	8494	8500	8506
71	8513	8519	8525	8531	8537	8543	8549	8555	8561	8567
72	8573	8579	8585	8591	8597	8603	8609	8615	8621	8627
73	8633	8639	8645	8651	8657	8663	8669	8675	8681	8686
74	8692	8698	8704	8710	8716	8722	8727	8733	8739	8745
75	8751	8756	8762	8768	8774	8779	8785	8791	8797	8802
76	8808	8814	8820	8825	8831	8837	8842	8848	8854	8859
77	8865	8871	8876	8882	8887	8893	8899	8904	8910	8915

N	0	1	2	3	4	5	6	7	8	9
78	8921	8927	8932	8938	8943	8949	8954	8960	8965	8971
79	8976	8982	8987	8993	8998	9004	9009	9015	9020	9025
80	9031	9036	9042	9047	9053	9058	9063	9069	9074	9079
81	9085	9090	9096	9101	9106	9112	9117	9122	9128	9133
82	9138	9143	9149	9154	9159	9165	9170	9175	9180	9186
83	9191	9196	9201	9206	9212	9217	9222	9227	9232	9238
84	9243	9248	9253	9258	9263	9269	9274	9279	9284	9289
85	9294	9299	9304	9309	9315	9320	9325	9330	9335	9340
86	9345	9350	9355	9360	9365	9370	9375	9380	9385	9390
87	9395	9400	9405	9410	9415	9420	9425	9430	9435	9440
88	9445	9450	9455	9460	9465	9469	9474	9479	9484	9489
89	9494	9499	9504	9509	9513	9518	9523	9528	9533	9538
90	9542	9547	9552	9557	9562	9566	9571	9576	9581	9586
91	9590	9595	9600	9605	9609	9614	9619	9624	9628	9633
92	9638	9643	9647	9652	9657	9661	9666	9671	9675	9680
93	9685	9689	9694	9699	9703	9708	9713	9717	9722	9727
94	9731	9736	9741	9745	9750	9754	9759	9763	9768	9773
95	9777	9782	9786	9791	9795	9800	9085	9809	9814	9818
96	9823	9827	9832	9836	9841	9845	9850	9854	9859	9863
97	9868	9872	9877	9881	9886	9890	9894	9899	9903	9908
98	9912	9917	9921	9926	9930	9934	9939	9943	9948	9952
99	9956	9961	9965	9969	9974	9978	9983	9987	9991	9996

Appendix 2

VALUES OF e^x AND e^{-x}

x	Function	0.00	0.02	0.04	0.06	0.08
0.0	e^x	1.0000	1.0202	1.0408	1.0618	1.0833
	e^{-x}	1.0000	0.9802	0.9608	0.9418	0.9231
0.5	e^x	1.6487	1.6820	1.7160	1.7507	1.7860
	e^{-x}	0.6065	0.5945	0.5827	0.5712	0.5599
1.0	e^x	2.7183	2.7732	2.8292	2.8864	2.9447
	e^{-x}	0.3679	0.3606	0.3535	0.3465	0.3396
1.5	e^x	4.4817	4.5722	4.6646	4.7588	4.8550
	e^{-x}	0.2231	0.2187	0.2144	0.2101	0.2060
2.0	e^x	7.3891	7.5383	7.6906	7.8460	8.0045
	e^{-x}	0.1353	0.1327	0.1300	0.1275	0.1249
2.5	e^x	12.182	12.429	12.680	12.936	13.197
	e^{-x}	0.0821	0.0805	0.0789	0.0773	0.0758
3.0	e^x	20.086	20.491	20.905	21.328	21.758
	e^{-x}	0.0498	0.0488	0.0478	0.0469	0.0460
3.5	e^x	33.115	33.784	34.467	35.163	35.874
	e^{-x}	0.0302	0.0296	0.0290	0.0284	0.0279
4.0	e^x	54.598	55.701	56.826	57.974	59.145
	e^{-x}	0.0183	0.0180	0.0176	0.0172	0.0169
4.5	e^x	90.017	91.836	93.691	95.583	97.514
	e^{-x}	0.0111	0.0109	0.0107	0.0105	0.0103
5.0	e^x	148.41	151.41	154.47	157.59	160.77
	e^{-x}	0.0067	0.0066	0.0065	0.0063	0.0062
5.5	e^x	244.69	249.64	254.68	259.82	265.07
	e^{-x}	0.0041	0.0040	0.0039	0.0038	0.0038

Appendix 3

INCOMPLETE GAMMA-FUNCTION RATIOS

u	$I(u,1)$	$I(u,2)$	$I(u,3)$	$I(u,5)$	$I(u,8)$	$I(u,10)$
0.0	0.0000	0.0000	0.0000	0.0000		
0.2	0.0332	0.0054	0.0008	0.0000	0.0000	0.0000
0.4	0.1107	0.0333	0.0091	0.0005	0.0000	0.0000
0.6	0.2087	0.0877	0.0338	0.0040	0.0001	0.0000
0.8	0.3124	0.1630	0.0788	0.0151	0.0009	0.0001
1.0	0.4131	0.2513	0.1429	0.0387	0.0038	0.0007
1.2	0.5058	0.3445	0.2213	0.0779	0.0117	0.0027
1.4	0.5885	0.4368	0.3081	0.1332	0.0279	0.0083
1.6	0.6605	0.5237	0.3975	0.2024	0.0558	0.0202
1.8	0.7219	0.6027	0.4848	0.2816	0.0973	0.0414
2.0	0.7737	0.6725	0.5665	0.3663	0.1528	0.0745
2.2	0.8169	0.7328	0.6406	0.4519	0.2204	0.1209
2.4	0.8525	0.7840	0.7058	0.5347	0.2973	0.1802
2.6	0.8817	0.8268	0.7619	0.6116	0.3796	0.2504
2.8	0.9054	0.8621	0.8094	0.6809	0.4631	0.3285
3.0	0.9247	0.8909	0.8488	0.7416	0.5443	0.4107
3.2	0.9402	0.9142	0.8811	0.7935	0.6204	0.4932
3.4	0.9526	0.9329	0.9072	0.8370	0.6892	0.5727
3.6	0.9625	0.9477	0.9281	0.8728	0.7498	0.6464
3.8	0.9705	0.9595	0.9446	0.9018	0.8016	0.7127
4.0	0.9767	0.9687	0.9576	0.9249	0.8450	0.7705
4.2	0.9817	0.9759	0.9677	0.9431	0.8805	0.8196
4.4	0.9857	0.9815	0.9756	0.9572	0.9090	0.8604
4.6	0.9888	0.9859	0.9816	0.9681	0.9316	0.8935
4.8	0.9912	0.9892	0.9862	0.9763	0.9491	0.9198

Appendix 4

VALUES OF THE STANDARD NORMAL DISTRIBUTION z

$$\int_{-\infty}^{z} \frac{1}{\sqrt{2\pi}} e^{-z^2/2} = p(z)$$

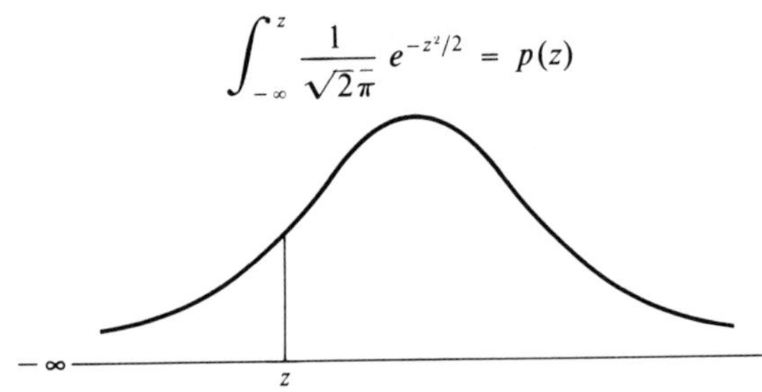

z	0	1	2	3	4	5	6	7	8	9
−2.9	0.0019	0.0018	0.0017	0.0017	0.0016	0.0016	0.0015	0.0015	0.0014	0.0014
−2.8	0.0026	0.0025	0.0024	0.0023	0.0023	0.0022	0.0021	0.0020	0.0020	0.0019
−2.7	0.0035	0.0034	0.0033	0.0032	0.0031	0.0030	0.0029	0.0028	0.0027	0.0026
−2.6	0.0047	0.0045	0.0044	0.0043	0.0041	0.0040	0.0039	0.0038	0.0037	0.0036
−2.5	0.0062	0.0060	0.0059	0.0057	0.0055	0.0054	0.0052	0.0051	0.0049	0.0048
−2.4	0.0082	0.0080	0.0078	0.0075	0.0073	0.0071	0.0069	0.0068	0.0066	0.0064
−2.3	0.0107	0.0104	0.0102	0.0099	0.0096	0.0094	0.0091	0.0089	0.0087	0.0084
−2.2	0.0139	0.0316	0.0132	0.0129	0.0126	0.0122	0.0119	0.0116	0.0113	0.0110
−2.1	0.0179	0.0174	0.0170	0.0166	0.0162	0.0158	0.0154	0.0150	0.0146	0.0143
−2.0	0.0228	0.0222	0.0217	0.0212	0.0207	0.0202	0.0197	0.0192	0.0188	0.0183
−1.9	0.0287	0.0281	0.0274	0.0268	0.0262	0.0256	0.0250	0.0244	0.0238	0.0233
−1.8	0.0359	0.0352	0.0344	0.0336	0.0329	0.0322	0.0314	0.0307	0.0300	0.0294
−1.7	0.0446	0.0436	0.0427	0.0418	0.0409	0.0401	0.0392	0.0384	0.0375	0.0367
−1.6	0.0548	0.0537	0.0526	0.0516	0.0505	0.0495	0.0485	0.0475	0.0465	0.0455
−1.5	0.0668	0.0655	0.0643	0.0630	0.0618	0.0606	0.0594	0.0582	0.0570	0.0559
−1.4	0.0808	0.0793	0.0778	0.0764	0.0749	0.0735	0.0722	0.0708	0.0694	0.0681
−1.3	0.0968	0.0951	0.0934	0.0918	0.0901	0.0885	0.0869	0.0853	0.0838	0.0823
−1.2	0.1151	0.1131	0.1112	0.1093	0.1075	0.1056	0.1038	0.1020	0.1003	0.0985
−1.1	0.1357	0.1335	0.1314	0.1292	0.1271	0.1251	0.1230	0.1210	0.1190	0.1170
−1.0	0.1587	0.1562	0.1539	0.1515	0.1492	0.1469	0.1446	0.1423	0.1401	0.1379

z	0	1	2	3	4	5	6	7	8	9
−.9	0.1841	0.1814	0.1788	0.1762	0.1736	0.1711	0.1685	0.1660	0.1635	0.1611
−.8	0.2119	0.2090	0.2061	0.2033	0.2005	0.1977	0.1949	0.1922	0.1894	0.1867
−.7	0.2420	0.2389	0.2358	0.2327	0.2297	0.2266	0.2236	0.2206	0.2177	0.2148
−.6	0.2743	0.2709	0.2676	0.2643	0.2611	0.2578	0.2546	0.2514	0.2483	0.2451
−.5	0.3085	0.3050	0.3015	0.2981	0.2946	0.2912	0.2877	0.2843	0.2810	0.2776
−.4	0.3446	0.3409	0.3372	0.3336	0.3300	0.3264	0.3228	0.3192	0.3156	0.3121
−.3	0.3821	0.3783	0.3745	0.3707	0.3669	0.3632	0.3594	0.3557	0.3520	0.3483
−.2	0.4207	0.4168	0.4129	0.4090	0.4052	0.4013	0.3974	0.3936	0.3897	0.3859
−.1	0.4602	0.4562	0.4522	0.4483	0.4443	0.4404	0.4364	0.4325	0.4286	0.4247
−.0	0.5000	0.4960	0.4920	0.4880	0.4840	0.4801	0.4761	0.4721	0.4681	0.4641

SOURCE: B. W. Lingdren and G. W. MacElrath, *Introduction to Probability and Statistics*, 2nd ed., The Macmillan Company, New York, 1966.

Appendix 5

TABLES OF t VALUES

Degrees of Freedom	Probability 0.50	0.10	0.05	0.02	0.01
1	1.000	6.34	12.71	31.82	63.66
2	0.816	2.92	4.30	6.96	9.92
3	0.765	2.35	3.18	4.54	5.84
4	0.741	2.13	2.78	3.75	4.60
5	0.727	2.02	2.57	3.36	4.03
6	0.718	1.94	2.45	3.14	3.71
7	0.711	1.90	2.36	3.00	3.50
8	0.706	1.86	2.31	2.90	3.36
9	0.703	1.83	2.26	2.82	3.25
10	0.700	1.81	2.23	2.76	3.17
11	0.697	1.80	2.20	2.72	3.11
12	0.695	1.78	2.18	2.68	3.06
13	0.694	1.77	2.16	2.65	3.01
14	0.692	1.76	2.14	2.62	2.98
15	0.691	1.75	2.13	2.60	2.95
16	0.690	1.75	2.12	2.58	2.92
17	0.689	1.74	2.11	2.57	2.90
18	0.688	1.73	2.10	2.55	2.88
19	0.688	1.73	2.09	2.54	2.86
20	0.687	1.72	2.09	2.53	2.84
21	0.686	1.72	2.08	2.52	2.83
22	0.686	1.72	2.07	2.51	2.82
23	0.685	1.71	2.07	2.50	2.81
24	0.685	1.71	2.06	2.49	2.80
25	0.684	1.71	2.06	2.48	2.79
26	0.684	1.71	2.06	2.48	2.78
27	0.684	1.70	2.05	2.47	2.77
28	0.683	1.70	2.05	2.47	2.76

Degrees of Freedom	Probability				
	0.50	0.10	0.05	0.02	0.01
29	0.683	1.70	2.04	2.46	2.76
30	0.683	1.70	2.04	2.46	2.75
35	0.682	1.69	2.03	2.44	2.72
40	0.681	1.68	2.02	2.42	2.71
45	0.680	1.68	2.02	2.41	2.69
50	0.679	1.68	2.01	2.40	2.68
60	0.678	1.67	2.00	2.39	2.66
70	0.678	1.67	2.00	2.38	2.65
80	0.677	1.66	1.99	2.38	2.64
90	0.677	1.66	1.99	2.37	2.63
100	0.677	1.66	1.98	2.36	2.63
125	0.676	1.66	1.98	2.36	2.62
150	0.676	1.66	1.98	2.35	2.61
200	0.675	1.65	1.97	2.35	2.60
300	0.675	1.65	1.97	2.34	2.59
400	0.675	1.65	1.97	2.34	2.59
500	0.674	1.65	1.96	2.33	2.59
1000	0.674	1.65	1.96	2.33	2.58
∞	0.674	1.64	1.96	2.33	2.58

SOURCE: R. A. Fisher and F. Yates, *Statistical Tables for Biological, Agricultural and Medical Research*, Oliver & Boyd, Ltd., Edinburgh.

Appendix 6

VALUES OF X^2

	Probability													
n	.99	.98	.95	.90	.80	.70	.50	.30	.20	.10	.05	.02	.01	.001
1	$.0^3157$	$.0^3628$	.00393	.0158	.0642	0.148	0.455	1.074	1.642	2.706	3.841	5.412	6.635	10.827
2	0.0201	0.0404	0.103	0.211	0.446	0.713	1.386	2.408	3.219	4.605	5.991	7.824	9.210	13.815
3	0.115	0.185	0.352	0.584	1.005	1.424	2.366	3.665	4.642	6.251	7.815	9.837	11.345	16.268
4	0.297	0.429	0.711	1.064	1.649	2.195	3.357	4.878	5.989	7.779	9.488	11.668	13.277	18.465
5	0.554	0.752	1.145	1.610	2.343	3.000	4.351	6.064	7.289	9.236	11.070	13.388	15.086	20.517
6	0.872	1.134	1.635	2.204	3.070	3.828	5.348	7.231	8.558	10.645	12.592	15.033	16.812	22.457
7	1.239	1.564	2.167	2.833	3.822	4.671	6.346	8.383	9.803	12.017	14.067	16.622	18.475	24.322
8	1.646	2.032	2.733	3.490	4.594	5.527	7.344	9.524	11.030	13.362	15.507	18.168	20.090	26.125
9	2.088	2.532	3.325	4.168	5.380	6.393	8.343	10.656	12.242	14.684	16.919	19.679	21.666	27.877
10	2.558	3.059	3.940	4.865	6.179	7.267	9.342	11.781	13.442	15.987	18.307	21.161	23.209	29.588
11	3.053	3.609	4.575	5.578	6.989	8.148	10.341	12.899	14.631	17.275	19.675	22.618	24.725	31.264
12	3.571	4.178	5.226	6.304	7.807	9.034	11.340	14.011	15.812	18.549	21.026	24.054	26.217	32.909
13	4.107	4.765	5.892	7.042	8.634	9.926	12.340	15.119	16.985	19.812	22.362	25.472	27.688	34.528
14	4.660	5.368	6.571	7.790	9.467	10.821	13.339	16.222	18.151	21.064	23.685	26.873	29.141	36.123
15	5.229	5.985	7.261	8.547	10.307	11.721	14.339	17.322	19.311	22.307	24.996	28.259	30.578	37.697

16	5.812	6.614	7.962	9.312	11.152	12.624	15.338	18.418	20.465	23.542	26.296	29.633	32.000	39.252
17	6.408	7.255	8.672	10.085	12.002	13.531	16.338	19.511	21.615	24.769	27.587	30.995	33.409	40.790
18	7.015	7.906	9.390	10.865	12.857	14.440	17.338	20.601	22.760	25.989	28.869	32.346	34.805	42.312
19	7.633	8.567	10.117	11.651	13.716	15.352	18.338	21.689	23.900	27.204	30.144	33.687	36.191	43.820
20	8.260	9.237	10.851	12.443	14.578	16.266	19.337	22.775	25.038	28.412	31.410	35.020	37.566	45.315
21	8.897	9.915	11.591	13.240	15.445	17.182	20.337	23.858	26.171	29.615	32.671	36.343	38.932	46.797
22	9.542	10.600	12.338	14.041	16.314	18.101	21.337	24.939	27.301	30.813	33.924	37.659	40.289	48.268
23	10.196	11.293	13.091	14.848	17.187	19.021	22.337	26.018	28.429	32.007	35.172	38.968	41.638	49.728
24	10.856	11.992	13.848	15.659	18.062	19.943	23.337	27.096	29.553	33.196	36.415	40.270	42.980	51.179
25	11.524	12.697	14.611	16.473	18.940	20.867	24.337	28.172	30.675	34.382	37.652	41.566	44.314	52.620
26	12.198	13.409	15.379	17.292	19.820	21.792	25.336	29.246	31.795	35.563	38.885	42.856	45.642	54.052
27	12.879	14.125	16.151	18.114	20.703	22.719	26.336	30.319	32.912	36.741	40.113	44.140	46.963	55.476
28	13.565	14.847	16.928	18.939	21.588	23.647	27.336	31.391	34.027	37.916	41.337	45.419	48.278	56.893
29	14.256	15.574	17.708	19.768	22.475	24.577	28.336	32.461	35.139	39.087	42.557	46.693	49.588	58.302
30	14.953	16.306	18.493	20.599	23.364	25.508	29.336	33.530	36.250	40.256	43.773	47.962	50.892	59.703

SOURCE: R. A. Fisher and F. Yates, *Statistical Tables for Biological, Agricultural and Medical Research*, Oliver & Boyd, Ltd., Edinburgh.

Appendix 7

TABLE OF F VALUES

0.01 probability of a larger value of F
ν_2 = degrees of freedom for numerator
ν_1 = degrees of freedom for denominator

ν_1 \ ν_2	1	2	3	4	5	10	16	20	30	40	100	∞
1	4050	5000	5400	5630	5770	6060	6170	6200	6260	6290	6340	6370
2	98.50	99.00	99.20	99.30	99.30	99.40	99.50	99.50	99.50	99.50	99.50	99.50
3	34.12	30.82	29.46	28.71	28.24	27.23	26.83	26.69	26.50	26.41	26.23	26.12
4	21.20	18.00	16.69	15.98	15.52	14.54	14.15	14.02	13.83	13.74	13.57	13.46
5	16.26	13.27	12.06	11.39	10.97	10.05	9.68	9.55	9.38	9.29	9.13	9.02
6	13.74	10.92	9.78	9.15	8.75	7.87	7.52	7.39	7.23	7.14	6.99	6.88
7	12.25	9.55	8.45	7.85	7.46	6.62	6.27	6.15	5.98	5.90	5.75	5.65
8	11.26	8.65	7.59	7.01	6.63	5.82	5.48	5.36	5.20	5.11	4.96	4.86
9	10.56	8.02	6.99	6.42	6.06	5.26	4.92	4.80	4.64	4.56	4.41	4.31
10	10.04	7.56	6.55	5.99	5.64	4.85	4.52	4.41	4.25	4.17	4.01	3.91
15	8.68	6.36	5.42	4.89	4.56	3.80	3.48	3.36	3.20	3.12	2.97	2.87
20	8.10	5.85	4.94	4.43	4.10	3.37	3.05	2.94	2.77	2.69	2.53	2.42
25	7.77	5.57	4.68	4.18	3.86	3.13	2.81	2.70	2.54	2.45	2.29	2.17
30	7.56	5.39	4.51	4.02	3.70	2.98	2.66	2.55	2.38	2.29	2.13	2.01
50	7.17	5.06	4.20	3.72	3.41	2.70	2.39	2.26	2.10	2.00	1.82	1.68
100	6.90	4.82	3.98	3.51	3.20	2.51	2.19	2.06	1.89	1.79	1.59	1.43
1000	6.66	4.62	3.80	3.34	3.04	2.34	2.01	1.89	1.71	1.61	1.38	1.11
∞	6.64	4.60	3.78	3.32	3.02	2.32	1.99	1.87	1.69	1.59	1.36	1.00

0.05 probability of a larger value of F
ν_2 = degrees of freedom for numerator
ν_1 = degrees of freedom for denominator

ν_1 \ ν_2	1	2	3	4	5	10	16	20	30	40	100	∞
1	160	200	220	230	230	250	250	250	250	250	250	250
2	18.51	19.00	19.16	19.25	19.30	19.39	19.42	19.44	19.46	19.47	19.49	19.50
3	10.13	9.55	9.28	9.12	9.01	8.78	8.69	8.66	8.62	8.60	8.56	8.53
4	7.71	6.94	6.59	6.39	6.26	5.96	5.84	5.80	5.74	5.71	5.66	5.63
5	6.61	5.79	5.41	5.19	5.05	4.74	4.60	4.56	4.50	4.46	4.40	4.36
6	5.99	5.14	4.76	4.53	4.39	4.06	3.92	3.87	3.81	3.77	3.71	3.67
7	5.59	4.74	4.35	4.12	3.97	3.63	3.49	3.44	3.38	3.34	3.28	3.23
8	5.32	4.46	4.07	3.84	3.69	3.34	3.20	3.15	3.08	3.05	2.98	2.93
9	5.12	4.26	3.86	3.63	3.48	3.13	2.98	2.93	2.86	2.82	2.76	2.71
10	4.96	4.10	3.71	3.48	3.33	2.97	2.82	2.77	2.70	2.67	2.59	2.54
15	4.54	3.68	3.29	3.06	2.90	2.55	2.39	2.33	2.25	2.21	2.12	2.07
20	4.35	3.49	3.10	2.87	2.71	2.35	2.18	2.12	2.04	1.99	1.90	1.84
25	4.24	3.38	2.99	2.76	2.60	2.24	2.06	2.00	1.92	1.87	1.77	1.71
30	4.17	3.32	2.92	2.69	2.53	2.16	1.99	1.93	1.84	1.79	1.69	1.62
50	4.03	3.18	2.79	2.56	2.40	2.02	1.85	1.78	1.69	1.63	1.52	1.44
100	3.94	3.09	2.70	2.46	2.30	1.92	1.75	1.68	1.57	1.51	1.39	1.28
1000	3.85	3.00	2.61	2.38	2.22	1.84	1.65	1.58	1.47	1.41	1.26	1.08
∞	3.84	2.99	2.60	2.37	2.21	1.83	1.64	1.57	1.46	1.40	1.24	1.00

Appendix 8

TABLE OF NATURAL LOGARITHMS

0.00–0.99

– 10 should be appended to each logarithm

N	.00	.01	.02	.03	.04	.05	.06	.07	.08	.09
0.0		5.395	6.088	6.493	6.781	7.004	7.187	7.341	7.474	7.592
0.1	7.697	7.793	7.880	7.960	8.034	8.103	8.167	8.228	8.285	8.339
0.2	8.391	8.439	8.486	8.530	8.573	8.164	8.653	8.691	8.727	8.762
0.3	8.796	8.829	8.861	8.891	8.921	8.950	8.978	9.006	9.032	9.058
0.4	9.084	9.108	9.132	9.156	9.179	9.201	9.223	9.245	9.266	9.287
0.5	9.307	9.327	9.346	9.365	9.384	9.402	9.420	9.438	9.455	9.472
0.6	9.489	9.506	9.522	9.538	9.554	9.569	9.584	9.600	9.614	9.629
0.7	9.643	9.658	9.671	9.685	9.699	9.712	9.726	9.739	9.752	9.764
0.8	9.777	9.789	9.802	9.814	9.826	9.837	9.849	9.861	9.872	9.883
0.9	9.895	9.906	9.917	9.927	9.938	9.949	9.959	9.970	9.980	9.990

For example, ln 0.48 = 9.266 – 10 = 0.734

n	ln *n*	*n*	ln *n*	*n*	ln *n*
1.0	0.0000	4.0	1.3863	7.0	1.9459
1.1	0.0953	4.1	1.4110	7.1	1.9601
1.2	0.1823	4.2	1.4351	7.2	1.9741
1.3	0.2624	4.3	1.4586	7.3	1.9879
1.4	0.3365	4.4	1.4816	7.4	2.0015
1.5	0.4055	4.5	1.5041	7.5	2.0149
1.6	0.4700	4.6	1.5261	7.6	2.0281
1.7	0.5306	4.7	1.5476	7.7	2.0412
1.8	0.5878	4.8	1.5686	7.8	2.0541
1.9	0.6419	4.9	1.5892	7.9	2.0669
2.0	0.6931	5.0	1.6094	8.0	2.0794
2.1	0.7419	5.1	1.6292	8.1	2.0919
2.2	0.7885	5.2	1.6487	8.2	2.1041
2.3	0.8329	5.3	1.6677	8.3	2.1163
2.4	0.8755	5.4	1.6864	8.4	2.1282

n	ln n	n	ln n	n	ln n
2.5	0.9163	5.5	1.7047	8.5	2.1401
2.6	0.9555	5.6	1.7228	8.6	2.1518
2.7	0.9933	5.7	1.7405	8.7	2.1633
2.8	1.0296	5.8	1.7579	8.8	2.1748
2.9	1.0647	5.9	1.7750	8.9	2.1861
3.0	1.0986	6.0	1.7918	9.0	2.1972
3.1	1.1314	6.1	1.8083	9.1	2.2083
3.2	1.1632	6.2	1.8245	9.2	2.2192
3.3	1.1939	6.3	1.8405	9.3	2.2300
3.4	1.2238	6.4	1.8563	9.4	2.2407
3.5	1.2528	6.5	1.8718	9.5	2.2513
3.6	1.2809	6.6	1.8871	9.6	2.2618
3.7	1.3083	6.7	1.9021	9.7	2.2721
3.8	1.3350	6.8	1.9169	9.8	2.2824
3.9	1.3610	6.9	1.9315	9.9	2.2925
				10.0	2.3026

Index